GARDEN PESTS & DISEASES *in* CANADA

The Good, the Bad and the Slimy

Rob Sproule

Lone Pine Publishing

© 2015 by Lone Pine Publishing
First printed in 2015   10 9 8 7 6 5 4 3 2 1
Printed in China

Lone Pine Publishing
87 East Pender Street
Vancouver, BC V6A 1S9
Canada

Website: www.lonepinepublishing.com

Library and Archives Canada Cataloguing in Publication

Sproule, Rob, 1978-, author
    Garden pests & diseases in Canada / Rob Sproule.

Includes index.
ISBN 978-1-55105-917-4 (pbk.)

    1. Garden pests--Canada. 2. Plant diseases--Canada. 3. Garden pests--Control--Canada.
4. Phytopathogenic microorganisms--Control--Canada. I. Title. II. Title: Garden pests and diseases of Canada.

SB603.5.S67 2015   635'.049   C2014-907411-5

Editorial Director: Nancy Foulds
Project Editor: Sheila Quinlan
Production Manager: Leslie Hung
Layout and Production: Volker Bodegom, Gregory Brown
Cover Design: Gerry Dotto
Front cover: foreground - Scarlet Lily Beetle © Shoshan67 / Thinkstock; background beetle - © SarahKellett / Thinkstock

Photography: AAFC 27; Alison Beck 187a; Alesha Braitenbach 142a; Olivier Bruchez (flickr) 102; Chicagoland Grows Inc. 38, 127b; Dohnal (Dreamstime) 177b; Tamara Eder 6, 12, 14, 17b, 18b, 19a&b, 20, 28, 35, 51b, 56, 75a, 82, 83, 89, 91, 108, 138b, 165a, 181a, 182; Jen Fafard 178; Derek Fell 7, 8, 9, 22, 39a, 74, 79a, 119a, 126b, 127a, 161, 162b; Erika Flatt 86a, 92a; Anne Gordon 44, 53, 103; Kai Hendry (flickr) 128; Rocky Houghtby (flickr) 115; Linda Kershaw 148; Liz Klose 99, 139, 173a, 186b; Debra Knapke 57; Matt Lavin (flickr) 151, 155, 157b; Mike Lewinski (flickr) 112a; Dawn Loewen 15a; Doug Macaulay 37, 50b, 79b; Heather Markham 55b; Tim Matheson 13b, 15b, 46, 107b, 111a, 113a&b, 114b, 116a&b, 119b, 134, 135, 138a, 144, 152, 154, 173b, 186a; Frank Mayfield (flickr) 153; Marilyn McAra 33, 96a&b, 97; NatureServe/Joseph O'Brien (flickr) 125; Scot Nelson (flickr) 100, 101a&b; Ninjatacoshell (Wikimedia) 118; Nuuuuuuuuuuul (flickr) 165b; Kim O'Leary 95; Lindy Oyama Bryan 73; Allison Penko 48b, 111b, 120, 184b; Laura Peters 23, 25, 58b, 62b, 63, 65a&b, 70, 71a,b,c&d, 75b, 85, 86b, 88a&b, 92b, 114a, 117, 121, 122, 123, 129b, 130, 147, 166, 167, 170, 180, 184a; photos.com 42; Rasbak (Wikimedia) 156, 157a; Robert Ritchie 30, 31, 36, 106, 107a; D. Gordon E. Robertson (Wikimedia) 150; Nanette Samol 13a, 39b, 59a&b, 60, 64a, 68, 98, 145, 177a, 181b, 183; Sabrina Setaro (flickr) 112b; Soil Foodweb Inc. 104a&b; Forest and Kim Starr (flickr) 164; Paul Swanson 67, 129a, 131a&b; TPI 41, 54; Mark Turner 45, 126a, 158, 159a&b; Sandy Weatherall 179; Gary Whyte 94a&b; Don Williamson 16, 105, 109, 110, 137b, 146, 160, 162a, 185, 187b; ZooFari (Wikimedia) 163.

Illustrations: Charity Briere 10, 18a, 62a, 90, 124, 133; Ivan Droujinin 17a, 21, 24, 26, 29a&b, 32, 34, 40a&b, 47, 50a, 51a, 52, 55a, 58a, 61, 64b, 84, 87, 143; George Penetrante 43a&b, 48a, 49a&b, 69, 93, 132; Gary Ross 168, 169a,b&c, 171, 172a&b, 174, 175a&b; Ian Sheldon 11a&b, 66, 72, 76, 77, 78, 80, 81a&b, 136, 137a, 140, 141a&b, 142b, 149, 176.

We acknowledge the financial support of the Government of Canada through the Canada Book Fund (CBF) for our publishing activities.

PC: 27

Table of Contents

Dedication

To Meg and Aidan, the guiding stars in my night sky.
You make everything possible.

Acknowledgements

First and foremost, thanks to Meg for her endless motivation and inspiration. Thank you to my family for all their support; I love you guys!

Thanks to the incredible staff at Salisbury Greenhouse, who were always there to answer questions and provide ideas. Your enthusiasm pushed me forward every day! Thanks in particular to Bob and Dave Sproule, Brett Kerley, Nancy Gillis, Kim Brown and Alison Beck.

Thanks also to all the gardeners who think that spiders are beautiful and strive to create an ecosystem in their backyard. The world is a little better because of you!

Introduction

Attitudes can be slow to change, but when it comes to pest control, they are changing. As early as a generation ago, gardeners were applying pesticides so liberally that many gardens were without any bugs at all. If a bug (any bug) popped its tiny head up, it was a declaration of war to which the home-owner responded with a full arsenal of chemical weaponry, dousing the yard with enough pesticides to make sure the plants could grow bug free.

Now, we're rethinking how we look at bugs. Instead of seeing them as one monolithic mass of creepy-crawliness representing everything bad in the garden, we're recognizing the incredible variety of the species, and what they do, in the garden. For every nasty bug that burrows into our birch leaves or sucks the sap out of our vines, there's another bug hunting it.

The most effective pest control has nothing to do with pesticides. It's about creating an environment where good bugs can live alongside the bad, fending them off and keeping their numbers at a manageable, stable rate. A healthy population of predators will keep the majority of pest infestations under control through the natural food chain.

Having a balanced ecosystem in the yard requires cutting back on chemicals. Pesticides are indiscriminate and kill everything in their path, good and bad alike. The problem with killing good bugs along with the bad is that the bad bugs bounce back much faster than the good bugs. Wipe out all the aphids and ladybugs, and in a couple of weeks, the aphid army will have returned with nary a ladybug in sight.

I wrote this book because, as a Canadian gardener, I've always been frustrated at the lack of practical guidance for chemical-free pest controls for the pests that actually affect Canadian gardens. I've seen an endless stream of people coming into the greenhouse looking for ways to deal with Canadian bugs. Increasingly, they're concerned about the health effects of chemicals and don't want their kids or pets running through a pesticide-laced yard. A poison is a poison is a poison—if you're a bug it's instant death; if you're a person it just takes longer.

HISTORY OF PESTICIDES

When you think of pesticides, I'm guessing you're conjuring mental images of big business, brightly coloured spray bottles lined up neatly on shelves, and workers draped in white protective gear amidst clouds of ominous vapour. While that pretty much sums up today's industry, pesticide use dates back thousands of years.

Agriculture as we know it started about 10,000 years ago, and people started worrying about plant pests about five minutes later. The first evidence of humans deliberately using compounds to control nasty critters is the Sumerians using sulphur about 4500 years ago. "Brimstone" bricks were burnt in fields to ward off insects and mites, often ritualistically as part of larger religious ceremonies.

It's fascinating that sulphur, which is the most commonly applied fungicide available today, was also the first one ever used. Readily available in nature and relatively safe to use, sulphur's tendency to appear in volcanic areas earned it ancient associations with hell (hence its name in antiquity, "brimstone.")

About 3000 years ago, the Chinese started using arsenic and mercury to control body lice. In retrospect, this was clearly a bad idea; however, arsenic use continued into the 1920s and 1930s, which is why people celebrated the onset of a new generation of pesticides (including the infamous DDT) in the 1950s. Everything is relative. While we now know DDT to be toxic, compared to arsenic, it must have seemed refreshingly safe.

Pyrethrum is another commonly found modern pesticide with ancient origins. Over 2000 years ago, Chinese farmers noticed that pests didn't seem to bother chrysanthemums and began using pyrethrum, which is derived from chrysanthemum flowers. Traders began to exchange pyrethrum, via chrysanthemum flowers, along the Silk Road, and it

Pyrethrum is found in several species related to chrysanthemums, including these daisies.

The selective herbicide 2,4-D revolutionized the way we kill dandelions.

found its way to European farms and gardens. During the Napoleonic wars, French soldiers used crushed mums to control fleas and body lice.

American farmers began using pyrethrum in 1860, and when World War I broke out, they began importing most of the mums they used for it from Japan. Today, it's the most widely used pesticide in North America and is often used on organic farms. Although it's considered the safest pesticide, it's not harmless. It kills beneficial insects, is toxic to fish and has been found to increase estrogen levels in humans. While it should be the first insecticide you reach for, try to exhaust all chemical-free methods first.

In 1939, a Swiss chemist named Paul Hermann Muller discovered the poisonous potential of a colourless, odourless, tasteless compound which would later be known as DDT. He won the Nobel Prize in Medicine for his discovery, and initially, DDT was used to great fanfare by farmers and to control populations of malarial mosquitoes.

Rachel Carson's seminal 1962 book *Silent Spring* exposed the carcinogenic truth of DDT and how it travels up the food chain to larger animals. Carson's book taught us that our bodies are more porous and absorbent of the chemicals around us than we thought, and she changed the way we think about, and use, pesticides.

In the 1940s, the introduction of the first selective herbicide (i.e., one that kills one group of plants but not another) revolutionized weed control worldwide. The chemical known as 2,4-D, which is still one of the most widely used herbicides in the world, kills broadleaf plants such as dandelions and thistles but not narrow-leaf plants such as wheat, corn and grass. Although it meets Health Canada's standards for health and safety (as of 2013), research is ongoing. 2,4-D is more toxic to pets than humans, given that they're smaller and spend more time on the lawn.

Today, the focus is thankfully less on reliance on pesticides and more on integrated pest management (IPM), which stresses biological controls as the first line of defence. IPM sounds technical, but it's really quite simple: it's about creating an ecosystem in your yard that's inviting to a wide variety of critters. A broad critter base in the garden is essentially a standing army of beneficial, predatory insects such as ladybugs, and other predators such as birds and spiders. Pest invasions are inevitable, but when they happen, your army mobilizes and usually gets them quickly under control.

LEVELS OF CONTROL

When we see a bug in the garden, our first reaction is often an emotional, knee-jerk instinct to kill it. If you remember nothing else from this book, please remember this: you don't have to kill everything. In fact, by not killing everything, you allow a world to form where the good guys begin to do the dirty work for you.

Passive control involves maintaining a healthy garden ecosystem that usually takes care of itself. Try to choose a wide variety of plants (native or northern climate species, if possible), and avoid using any pesticides. Strive to have mature trees where birds can perch, hiding places for spiders to lurk, and plenty of flowers to attract bees. Strong, vigorous plants repel an astonishing array of pests and diseases, so keep your yard bountiful and your predators numerous.

If you break down and use chemicals out of desperation, you will need to rebuild your entire army of beneficial insects. Spraying a pesticide such as pyrethrum will kill more than aphids; it will also kill all the ladybugs that eat the aphids. Once the pyrethrum has cleared (a few days), the aphids will bounce back quickly and in numbers that will be harder for any surviving ladybugs to control. Your passive controls will be overwhelmed, and you'll need to nuke the yard yet again. This is the disturbing way in which we come to rely on chemicals.

When the outbreak happens so quickly your army can't repel them (and it does

It is futile to spray aphids with pesticides.

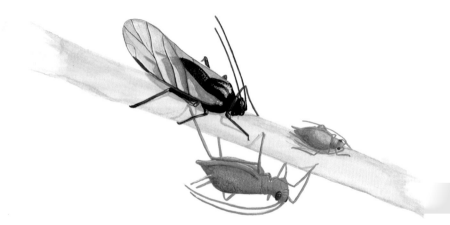

Ladybugs are one of the good guys in the garden.

happen as part of natural cycles), it's time to roll up your sleeves. There are active controls that are still earth-friendly. You may need to break out the hose to wash off those aphids, or pour boiling water on that nasty hill of stinging red ants. Often these tactics don't require that you buy any additional supplements from the store, but sometimes you'll need to.

Not every bottle on the chemical shelf at your local garden centre is dangerous. Products such as insecticidal soap (which is basically just dish soap) and sulphur dust are benign enough to use without fear of side-effects. Always look past the brand name and check which active ingredient is listed on the bottle. If you're not familiar with it, ask a staff member about its effectiveness and toxicity.

The future of pest control is in biologicals—predators—such as BTK (*Bacillus thuringiensis* var. *kurstaki*, a bacteria safe for humans but deadly for pests), nematodes, parasitic wasps and even ladybugs, all of which are available for purchase at large garden

centres. My commercial greenhouse made the switch from pesticides to biological controls in 2012, and the results have been amazing. Not only do we have far fewer pest bugs, but also, other beneficial creatures, such as orb spiders, are thriving.

Parasitic wasps are becoming more available for purchase as word of their effectiveness gets out.

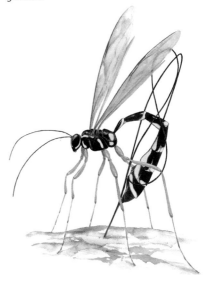

A note about using pest control products: treat the product, no matter its toxicity, like medicine. The dosage is always written clearly, sometimes in terms of mixing and sometimes in terms of frequency. A surprising number of people seem to think that if they just use more of it, the bugs will go away faster. This thinking is just plain dangerous. The dosage amounts aren't random; they are what Health Canada considers a safe level of exposure. Exceeding these amounts is unsafe to you and your pets, and it won't make that aphid any deader than it would have been at the proper dosage.

Unfortunately, many people choose chemical controls first, including 2,4-D, pyrethrum, etc. In a healthy, well-balanced yard, you should almost never have to resort to chemicals, but in unhealthy yards, you almost always have to because you've killed your own army.

CREATING A HEALTHY YARD

If I could manufacture a well-balanced yard spray and bottle it, I'd make millions. A healthy backyard ecosystem is far and away the best pest and disease control there is. The healthier plants are, the better they can repel diseases, and the resident army of beneficial insects repels all but the most virulent of pest invasions.

Creating a healthy yard takes a little time, but once established, it becomes a virtually self-sustaining ecosystem with more bountiful fruit, more flowers and chemical-free grass that you'll want your kids to run through barefoot. Here's how to do it.

SOIL

Soil is every healthy yard's foundation. It's the beginning of the food chain for both plants and animals—it's where abundance starts. Your yard's soil is an ecosystem unto itself, from microscopic

Creating a healthy, self-sustaining backyard ecosystem takes time and effort, but is worth it.

bacteria, fungi and nematodes to insects and mites to earthworms, all jostling for resources. Not only do soil-borne organisms create and unlock nutrients for plants to take up, but they also make tasty snacks for larger critters.

The best part about creating healthy soil is that you don't really have to do anything. Don't douse it with chemicals, avoid digging unless necessary, and the microbial universe will unfold by itself.

If your soil is very depleted or overly used (as is common with vegetable gardens), an annual top-dressing of compost will keep the dirt healthy. Layer on 5–10 cm every fall or spring, and the microbes will quickly colonize downward, rejuvenating the soil.

Use compost (below) to rejuvenate your soil (above).

Plants

Healthy gardens always have one thing in common: diversity. When you're choosing your plants, imagine that you're casting a play. Every player brings a different style and different gifts to the overall performance. The broader the array of players, the longer the audience is going to stand up and cheer.

Take a look at your lawn. While lawn is useful for picnics, pets and playing with the kids, most people have more than they need. Ecologically, lawn is a desert. It's not diverse and encourages us to use chemicals such as inorganic fertilizer and 2,4-D. Look at the lawn you don't use and imagine a lush perennial bed there instead, replete with summer bloomers and teeming with butterflies and bees. Now go rent that sod-stripper!

Opt for native or other northern-climate plants whenever possible. Not only will they require less winterizing and watering than more tender exotics, but having evolved in the same climate in which you're now growing them, they will have built-in pest and disease control. Buy a hardy shrub, plant it, and your path to passive pest resistance has started.

Choose as wide a diversity of plants as possible—annuals in containers, perennials in borders, veggies in a garden plot. If you have a wet spot in the yard, see it as an opportunity for more diversity and purchase some water-loving plants. If you have a hot spot, revel in that and build a succulent garden.

Trees are a vital part of any backyard ecosystem. Not everyone has the space for a giant American elm, but most people have space for an apple tree. Berry producers, such as mountain ash, will attract birds throughout winter and early spring that are looking to feast on the fermenting berries.

Include a diversity of plants in your perennial beds.

Berry trees will attract birds to your yard in winter (above); use insect traps to monitor the numbers and types of bugs in your garden (below).

Insects

Healthy yards teem with life of all kinds, including insects. While the "just kill it" mentality holds that all insects are bad, the reality is that only a small percentage of them are actual pests. The rest are either beneficial or neutral to humans and our plants.

Balance equals stability. A healthy yard always needs to have a few aphids in it because if it didn't, the resident ladybugs would either starve or fly away, leaving the yard helpless when the aphid invasion does happen. A stable population of all critters prevents the exponential explosion of one species.

Live and let live as much as you can. If there's an anthill tucked in the back and it's not hurting anything, let it be. Trying to micromanage an ecosystem always backfires. Our role is to provide a diverse, healthy setting, watch for anomalies that will require our attention and, barring those, interfere as little as possible.

A bird feeder will attract birds, which will then also eat a few insects.

BIGGER CRITTERS

Moving up in the food chain, big critters are the icing on the cake for a well-balanced yard. Resident animals such as birds will help disperse seeds, eat insect pests and add...ahem... fertilizer to the garden.

Animals need a reason to visit your yard, whether that's food, water, a hiding spot or long-time housing prospects. Providing a bird bath, having some seed available (keep the feeder a little ways from the house to keep mice away from your foundation) and planting bushes and trees with edible berries will serve as welcome signs.

BUGS ON THE MOVE

Fortunately, as it is with humans, plants tend to develop an immunity, over thousands of arduous years, to the pests and diseases they have evolved with and are accustomed to. Although that's not to say that outbreaks aren't still costly, the plants' natural defences

go a long way to minimize damage. Dutch elm disease, for example, hails from China, but the elms there, having evolved natural deterrents against elm bark beetles, tend to be unaffected.

Unfortunately, as it is with humans, bugs move. In this book, you'll find beetles hitching rides on cargo ships, caterpillars jumping on camping trailers and fungal spores drifting on breezes across entire towns. Through their own volition, human help and climate change, which is making Canada ever more hospitable to nasty creepies of every type, the bugs are coming.

In this book, I'll try to tell you where that particular bug comes from and how it got to Canadian shores. Sometimes it's surprising (did you know that Canada has no native earthworms?), and sometimes downright terrifying (the emerald ash borer chewing its way westward).

Canadian boulevard trees are an excellent example of how invasive pests could dramatically alter the landscape of our cities. Dutch elm disease continues to be an imminent threat, having wiped out millions of stately elm trees already. Ash trees, which largely replaced elms, are under threat from the emerald ash borer. And now aspen trees are also under threat from bronze leaf disease, leaving very few options remaining.

The nasties in this book are not just troublesome garden pests. Some of them are clear and present dangers to crops and our urban standard of living, and governments across Canada are scrambling to contain the spread.

The emerald ash borer (above) is a major threat to our ash trees and is making its way west. Dutch elm disease is threatening elms along boulevards across the country (below).

Aphids

Here in Canada, we're pretty lucky when it comes to insects—we don't have termites that eat houses or army ants that eat everything—but we do have aphids. Also known as plant lice, the lowly aphid (Aphididae) ranks up there with the mosquito as most loathed local insect.

IDENTIFICATION

In most ways, aphids are very easy to spot. They are pear shaped and always appear in clusters, grouping in dense masses on the tips of new plant growth or on the undersides of leaves. They love heavily fertilized plants because there are lots of yummy new leaves.

In other ways, they're tricksters. Numerous species are native to Canada, and they can be black, green, grey, spotted or even furry. If you want a reminder of what they look like, check your nearest petunia in September.

LIFE CYCLE

Aphids have a fascinating life cycle. They are the most remarkably adapted little buggers I've ever seen.

Aphids reproduce like crazy, both sexually and asexually. In spring, over-wintered eggs hatch into females that are already pregnant with thousands of young.

Until fall, aphids are all female, and a single female can hatch dozens of generations in one season. The nymphs mature in a week and start their own "families." It's an exponential curve that quickly adds up to millions of critters. Some species even give birth to females that are already pregnant!

In fall, the females start hatching both females and males. Some of them then sprout wings and fly into the familiar black swarms that we all hate in late August and September.

They overwinter by attaching their eggs to plants. When winter winds kill off the adults, the eggs persist, awaiting spring.

DAMAGE

Aphids are sap suckers and, because they are exceedingly good at it, are very destructive. Depending on the afflicted plant, you may notice yellow spots or wilted or curled leaves.

They secrete a sticky, tell-tale "honeydew." The secretion can make it easier for mould to set in. Aphid honeydew also often attracts ants, which "farm" the aphids by stroking their bodies with their antennae to collect the honeydew.

CONTROL

Luckily, aphids are the twinkies of the insect world. They have almost no physical defences, and almost everything likes to eat them. The best way to control them is to encourage predators.

Releasing ladybugs is a great way to control aphid populations. As long as there is a juicy buffet in the garden, the ladybugs should hang around. If ladybugs don't work, try applying some neem oil.

If you start early, you shouldn't need to resort to chemicals. Spraying the undersides of the leaves with a strong jet of water will usually knock the bugs off. If the numbers are getting out of control, try spraying insecticidal soap. You'll need to repeat spraying a few times to stop their life cycles. In late fall, spray your plants down with a strong spray of water to knock the eggs off so they don't overwinter and hatch into a new menace the next spring.

If you resort to pyrethrum, which is the active ingredient in many pesticides, to control your aphids, be warned that pyrethrum will wipe out the good guys as well is the bad guys, and the much simpler aphids will bounce back a lot faster. In other words, using harsh chemicals only makes the situation worse in the long run.

Keep aphids off your other plants by growing calendula; aphids love it, yet it will continue to bloom even with a heavy infestation.

Apple Maggot

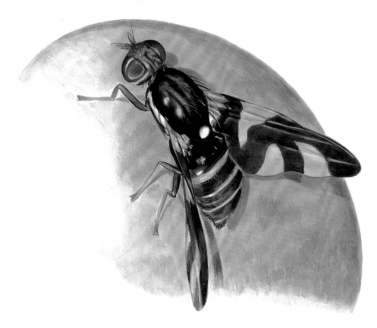

We've all heard murmurs about apple maggots (*Rhagoletis pomonella*) and how they pose a real threat to backyard apple production across Canada. I'd like to say that the stories are hyped up and the maggots are not that much of a threat, but unfortunately they are.

Also called railroad flies, apple maggots have been a century in Canada. Given the extent of the damage they've done to commercial and residential apple crops in B.C., it's important to understand the gravity of the threat, not just to backyard gardens but to commercial orchards, as well.

IDENTIFICATION

You'll be able to spot the flies from their tell-tale black and white striped wings; think of them as flying raccoons. They will spread outward and can fly in a 200–300 m radius looking for our lovely apple trees to infect and devour. It's not just apples at risk; they can strike hawthorns and even sour cherries, plums and pears.

LIFE CYCLE

The pupae stay tucked under the soil surface throughout winter and emerge as flies, usually after a good rain to soften the earth, in late June or early July. While the flies don't do any damage themselves, they buzz upwards and lay their eggs in the apples, which hatch into ravenous larvae.

Pick ripe apples as soon as they fall to break the apple maggots' life cycle.

After a summer spent gorging on your apples, the fat maggots are ready to overwinter. When the ripe apples fall, they crawl out of the apple and into the insulated earth, which will serve as their winter home.

DAMAGE

Apple maggots will ruin your apples. While the maggoty apples are, technically, edible, the thought of eating maggots is unpleasant at best. They can be fed to livestock (from a trough, not the ground), or even used to make cider or applesauce.

CONTROL

You can hang apple maggot lures and traps in your trees in early summer to snag the roving flies. The trap is basically a poisoned apple (Snow White–style), only it's coated in sticky pheromones instead of poison. When the randy flies, looking to make more maggot babies, land on the apple, they are stuck to it forever. It's romantic in a dark and twisted way.

The best way to control apple maggots is to break their life cycle. Their most vulnerable stage is the few days that elapse between the apple falling and them burrowing into the ground for safety. If you pick up the apples promptly, you can prevent the maggots from getting established underground or even break an infestation in one life cycle.

Another method is to lay a tarp down around the tree, extending from the trunk to the drip-line (outer edge of the leaves). Unless you want to kill the lawn underneath, only do this while the apples are falling. The maggots' attempts to burrow underground will be foiled, and you'll be able to pick the apples up in one go instead of having to monitor the tree daily.

Put infested apples in the garbage instead of the compost. It may seem like a waste, but otherwise the maggots will simply burrow into the compost as if it were the ground and continue their life cycle.

Prevention is a chore, so as you do it, make sure to knock on the door of any neighbours with apple trees. As they probably won't be aware of the situation, explain it to them and ask if they would please contribute by picking up their apples. It takes a neighbourhood to keep them at bay and/or eradicate them.

While apple maggots are a serious threat, they aren't unstoppable. The most important step to stopping them is to educate yourself and your neighbours about the problem and the steps to take if they have apple trees.

Monitor your apple trees; prevention is key to a healthy orchard.

Cottony Psyllid

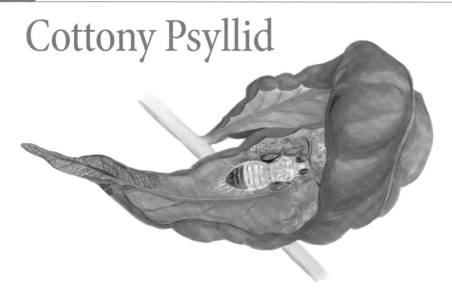

With the intensely cold winters we endure on the prairies, we have precious few options when it comes to statement-making shade trees. As such, it's understandably frustrating when a tiny European upstart starts messing with our ash trees.

Cottony psyllids (*Psyllopsis discrepans*), alias jumping tree lice, leapt across the pond and are showing up in Alberta, Saskatchewan, North Dakota, Minnesota and more. Most of us have seen the once-proud ash leaves curled up into a sad little cauliflowered clump.

Foresters on the prairies have cut and burned tens of thousands of trees in an effort to control the pest, which otherwise has no Canadian predators. Compounding the problem is that their habit of living inside folded leaves makes them devilishly difficult to hit with contact sprays.

IDENTIFICATION

They don't get easier to spot than the cottony psyllid. The nymphs both suck the juices out of and inject toxins into black ash and Manchurian ash leaves (they don't attack green ash), causing the leaves to collapse inward.

As the leaf curls into itself, the critters spin themselves a cottony cocoon to pupate in. The 3 mm long, yellow, aphid-looking insect that emerges doesn't do further damage to the tree.

LIFE CYCLE

Little is known about cottony psyllid biology, but if it's like similar critters, it emerges daisy fresh in spring to wreak havoc on its chosen victim. The immature nymphs, which attack as soon as the leaves unfurl, do the damage as they transform the leaves into cottony condos.

After pupating, the tiny pests are rarely seen, and we know little about their behaviour.

DAMAGE

The good news is that the psyllids won't usually kill your tree directly. What they will do is take a little wind out of your boastful shade tree's sails, as your tree struggles to pull in enough nutrients via crippled leaves to survive winter.

The cosmetic damage has a dark side, though, in that the reduced amount of nutrients the tree can take in weakens its entire system, making it susceptible to both early leaf drop and other, more serious pests and diseases.

If left untreated, the insects will spread to more leaves and, eventually, may weaken the tree to the point that it can't pull in enough nutrients to survive.

CONTROL

Cottony psyllids are vulnerable for only a short time between when they hatch and when they curl the leaf around them. You'll need to apply insecticidal soap after bud-break but before the leaves have been significantly curled. After that, the insects are either safely ensconced or flying about, and sprays such as insecticidal soap won't touch them.

The very short window (about one week), combined with the unfortunate fact that the damage is usually high in the canopy, makes spraying them a tricky and exacting science. No spray is effective outside of this short window.

Other than spraying, make sure to water your ash well. Healthy trees can often fight off minor infestations on their own, so as soon as the first leaves curl, grab the hose and give your tree's immune system a needed boost.

If a tree becomes completely infested, it will need to be removed and, in order to eradicate the pest fully, burned.

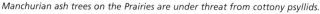

Manchurian ash trees on the Prairies are under threat from cottony psyllids.

Emerald Ash Borer

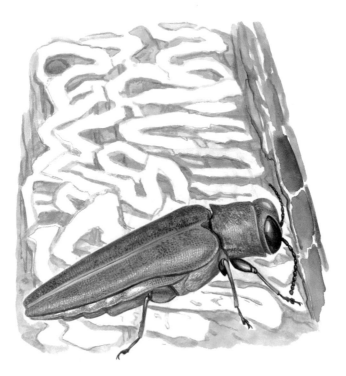

First it was the Mongol hordes, then it was the Red Army; now there's a new threat approaching from the East. The emerald ash borer (*Agrilus planipennis*), native to Asia, is highly destructive and sweeping westward at an alarming rate.

It made landfall in North America in 2002 by stowing away on Asian cargo ships that docked in Detroit. As of fall 2013, thanks to its ability to fly up a kilometre away to find a new host tree, it has spread to 22 states and two provinces (Ontario and Quebec), where it's already killed hundreds of millions of ash trees.

The emerald ash borer has no natural enemies in Canada and kills 99 percent of the trees it infects within six years as several generations munch through them. Western Canadian cities, with their heavy reliance on ash for urban forests, are justifiably spooked. Note that it only attacks the genus *Fraxinus*, and it ignores the unrelated mountain ash.

IDENTIFICATION

For big tree lovers, the emerald ash borer is one of the scarier beetles out there, and we all need to be educated about what to watch for. Unfortunately, early infestation is hard to spot. Ash trees will look healthy even as they're being devoured from the inside out. If you were to peel the outer bark away, you'd see the zigzagging tunnels that feasting larvae leave behind. You may be able to save your tree if you catch the following subtle early signs. More importantly, you will help stop the spread to other trees around it and could slow an infestation.

Larvae grow to almost 3 cm long before pupating. You won't see them, but watch for vertical cracks in the outer bark, which often appear as the larvae move underneath.

Trees under stress will often send up more suckers than usual. If you notice a lot of suckers growing from the tree's base, and you're in an infected area, alert your urban forestry officials.

Watch for woodpeckers. One of this borer's few North American enemies, they are very good at finding larvae and may favour infested trees.

Lastly, watch for 3–4 mm wide, D-shaped exit holes in spring. These appear only after a year or more of infestation, but at least you'll know if the tree is infected.

Adults are almost 1 cm long, bullet-shaped, and a striking, rich green. If you see them in the canopy of an ash tree, you've got trouble.

LIFE CYCLE

Adults spend their lives munching on leaves in the tree canopy and laying 1 mm wide, creamy white eggs amidst crevices in the rough bark.

The larvae, once hatched, chew into the tree's inner bark to munch the year away. They eat voraciously through layers of xylem (think of xylem as the tree's blood vessels), growing and chewing wide tunnels throughout the tree before overwintering inside.

After pupating, the adult beetle emerges in spring to either feast on the leaves of the same tree or fly up to a kilometre away to find its next victim.

The life cycle of this beetle can take one to two years to complete, depending on the trees' health and weather conditions. Sickly trees enduring drought years literally get chewed up the fastest.

Urban forests are under threat from the emerald ash borer.

DAMAGE

While the adults' damage is only cosmetic, it's often the wide-scale chewing of the crown that identifies the beetles' presence because the larvae are hidden within inflicted mortal wounds. By the time the crown is chewed back, tree death is usually inevitable. The tunnels the larvae chew throughout the tree destroy the tree's ability to transport water and nutrients through its system, eventually killing it.

CONTROL

The best way to fight this beetle is to keep your ash trees healthy and well pruned. Emerald ash borers establish themselves in dead or dying branches first and infect the rest of the tree from there. If you have ash trees you've been neglecting, get out there to prune and water them to begin the long process of building up their formidable immune system.

Other control methods, including systemic injections and biological (predatory) controls, are currently being researched as North American governments grapple with how to control the emerald borer when it inevitably arrives on their doorstep.

Firewood is the most common way to spread the beetle, and many infested areas prohibit ash firewood from leaving. Be wary of transporting ash firewood, wherever you are, or you may inadvertently spark an outbreak.

Well-pruned ash trees are the best defence against this pest.

Japanese Beetle

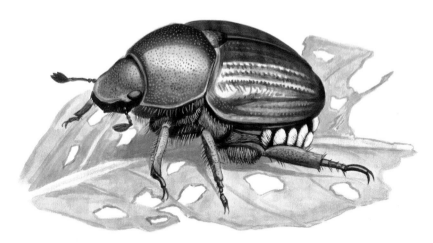

Throughout the eastern half of Canada, the name Japanese beetle (*Popillia japonica*) strikes fear in gardeners' hearts. Native to Japan, it stowed away to New Jersey via a shipment of iris bulbs in 1912. In 1939, it snuck into Nova Scotia via a ferry from Maine, and we've been fighting it ever since.

In Japan, where predators have evolved alongside it, beetle populations rarely spike beyond the reach of natural controls. North America doesn't enjoy such predators, and the invasive pest ravages commercial crops and residential gardens alike wherever it goes.

IDENTIFICATION

Japanese beetle larvae live under-ground. You won't see them, but you will see irregular brown spots in your lawn. After that, watch for birds taking interest in those areas and for raccoons, skunks and other opportunistic critters digging up your lawn for a squishy treat.

Adult beetles are about 1.25 cm long and have tell-tale iridescent copper plates on their back. Look for skeletonized leaves wherein the beetle has devoured the tender bits, leaving only the leaf's bare bones behind. Severely infested trees can look like they've been torched.

Japanese beetles are attracted to rose bushes...

LIFE CYCLE

These beetles spend the majority of their lives (10 out of 12 months) as grubs under the soil, where they delight in feeding on grass roots in late summer and fall and again in spring before the adults emerge.

After emerging from the soil in late June or early July, the groups of adults spend a week munching on low-growing plants before graduating up to fruit and shade trees. They're sunbathers and are most active on hot, sun-drenched days. You'll get a reprieve during cold, cloudy spells.

At summer's end, the adults lay their eggs in the ground. The eggs soon hatch into nasty grubs to begin the cycle anew.

DAMAGE

The Japanese beetle is two pests in one, with both the larvae and the adults doing significant and unique damage. They're voracious and are a serious hazard to any plant they sink their teeth into.

Adult Japanese beetles have a broad range of victims, feasting on elm, maple, rose, zinnia, corn, asparagus, grape, apple, blueberry, blackberry and many more for a baffling total of over 300 species.

CONTROL

The Canadian government takes Japanese beetle containment seriously enough to have established a regulated area containing the infected provinces. Plants and soil can be moved around within the regulated area (which includes Ontario, Quebec, New Brunswick, Nova Scotia and P.E.I.) but not outside of it.

Although many creatures like to snoop around for the grubs, the absence of natural predators for the imported

...and will do severe damage if not controlled early.

beetle has made natural controls very difficult. You can hang specially designed pheromone-enriched traps around the yard, but be warned that you may end up attracting lusty beetles from outside your garden, as well.

Control is very difficult once they're established, so early detection and prevention is key. Watch your lawn in spring, and if you see the tell-tale signs, head to a large garden centre for some BTK (*Bacillus thuringiensis* var. *kurstaki*). Apply this naturally occurring bacteria (which will only attack the grubs) to all affected areas.

Japanese beetles don't eat all plants, although it can sometimes seem that way.

They loathe catnip, chives, garlic and tansy so much that the plants are effective natural deterrents when planted around your grapes, blueberries and other crops you want to save.

If you're adventurous, the internet is full of home remedies. One gardener suggests sprinkling baby powder on the leaves the beetles want to eat, while another advocates sucking them up with a leaf blower switched to "suck" (though this may damage the plant as much as the beetle). The only home remedy that I can vouch for is to mix a few tablespoons of dish soap to a pail full of water and spray it directly onto the critters.

Scarlet Lily Beetle

Whether it's their ease of growth or stunning flowers or just that we're surprised that such a gorgeous plant can survive here, lilies are many Canadians' favourite flower. They are the full-throated operatic prima donnas of the perennial garden, and they're under threat by an unwelcome intruder.

The scarlet lily beetle (*Lilioceris lilii*), alias red lily beetle or lily leaf beetle, is a European native that hitched a ride to Canada on lily bulbs in the mid 1940s. Over the decades, it's settled westward as far as Alberta.

IDENTIFICATION

Vigilant gardeners (as most lily-lovers are) will first notice brown, sticky frass clumping on the leaves' undersides. Those same leaves will be entirely chewed up shortly after.

The adult beetles are an unmistakable bright red with black undersides; they look like futuristic plant-eating robots. They can be confused with spotless ladybugs, but they are more narrow and just plain meaner-looking than the beneficial insects.

LIFE CYCLE

Lily beetles spend winter quite comfortably cuddled into your perennial bed soil. After waking up in spring, they don't take long to start reproducing, with females laying up to 450 eggs per season in batches of a dozen. The yellowish orange eggs hide under lily leaves, usually tucked against the midrib.

The larvae are voracious and spend a few weeks chomping lilies while covered in a pile of their own excrement (if it's to camouflage as bird droppings, it's very convincing). Once they've eaten their fill of your lilies, they drop to the ground and pupate in a cocoon of dirt and saliva—enchanting creatures.

DAMAGE

Feasting on all species of lilies and fritillarias (though not daylilies), lily beetles are quickly chewing their way into Canadian gardening infamy. They're virtuosos at decimating lily collections that have been amassed over decades in record time, devouring leaves, stems, flowers and all.

CONTROL

Although there are plenty of parasitic wasps in Europe to feed on them, lily beetles have no natural enemies in North America. As such, they spread rapidly and will move quickly from ornamental plants to wild lilies, where they can do massive damage.

Before you plant your spring lily bulbs, inspect them for anything suspicious. This is their favourite mode of getting into your yard.

The most effective control is also the most laborious; inspect your lilies regularly, and if you find any lily beetles (either adults or larvae), pick them off manually. If you can't stomach crushing them, drop them in a bucket of soapy water.

Sprinkling diatomaceous earth on the larvae will dry them up, but you'll need to hit the leaves' undersides to get to the pests.

If you're persistent (to the tune of every five to seven days), neem oil will kill young larvae. Potent chemicals such as malathion will work, but I recommend that you stay away from them; lilies are a favourite haunt for bees, and strong chemicals could be fatal to them

and other beneficial critters in your garden.

Be forewarned that, when threatened, lily beetles will often play dead. As you close in on one with your pincers, expect it drop to the ground, put its legs in the air and wait for you to go away. Squish it anyway. They will also emit a high-pitched squeak as you get close, which can be fairly unnerving.

Inspect your lilies regularly for scarlet lily beetles.

Gypsy Moth

In 1869, a French naturalist attempted to pioneer a North American silk industry. His alchemy included breeding European gypsy moths (*Lymantria dispar dispar*) with North American silk worms, but something went wrong. A few insects escaped and quickly established themselves in the northeastern U.S. and eastern Canada. The gypsy moth's reign of terror had begun.

Gypsy moths earned their name from their tendency to travel long distances and, in doing so, spread like wildfire. To make matters worse, the Asia gypsy moth (*L. dispar asiatica*) has recently established a beachhead in Vancouver, opening a second front in the war against an invasive, destructive enemy.

Since 1970, gypsy moths have defoliated about 325,000 square kilometres of forest in North America and continue to spread. Governments at all levels have taken notice.

Gypsy moths especially favour oak trees.

IDENTIFICATION

The voracious caterpillars (larvae) hatch at 3 mm long in spring but gorge themselves to almost 90 mm before pupating. Young caterpillars are dark brown and fairly plain looking. As they grow, they develop deep bumps across their back and sprout rows of coarse black hairs. Rows of blue and red dots along their back make them very recognizable.

Adult moths' appearance varies by sex. Male moths are mottled brown and about 20 mm long. The flightless females are white with brown highlights and measure about 30 mm long.

LIFE CYCLE

Gypsy moths spend most of the year as eggs, hatching into larvae in spring. The caterpillars do massive damage, feeding until they pupate in early to midsummer into the moth stage. The moths exist for a single purpose: to lay eggs and spread more munching mayhem.

Each fuzzy egg mass contains 100 to 1000 eggs. The bigger the egg mass (they range from dime size to bigger than a loonie), the worse the overall infestation will be.

Gypsy moths really get around. When first hatched, wind can carry the tiny larvae up to a kilometre away. The caterpillars favour oak but will feed on almost 500 species of trees, both deciduous and coniferous. By the time they pupate in early July, each caterpillar has eaten about 1 square metre of foliage, and the damage is done.

Even one caterpillar can do a lot of damage to a tree.

DAMAGE

No tree is safe in a gypsy moth–infested forest. Stripped trees are left severely weakened, sometimes beyond recovery. Like most pests, they don't differentiate between forests and yards, and they will happily leap into your ornamental trees whenever they can.

CONTROL

The best way to control gypsy moths is to destroy the egg masses before they hatch. The best time to do this is between November and April. Strap on the snowshoes and venture out to explore the trees around you. You're looking for fuzzy, tan-coloured masses clinging to the sides of trees, buildings and even outdoor furniture; you'll know them when you see them. With a dull knife, scrape the mass into a bucket of hot water.

They will also lay egg masses on trailers, trucks, quads and boats so that summer campers can drive them up to hundreds of kilometres away to start a new infestation the following spring. Before pulling away from your campsite, put a quick egg mass check on the agenda.

If you miss the window for eggs and the caterpillars hatch, the job just got harder. You can either handpick them (tough if they're in tall trees), or you can use BTK (*Bacillus thuringiensis* var. *kurstaki*), which is a naturally occurring bacteria that kills caterpillars within five days while being safe for other critters.

A small parasitic wasp was introduced to eastern North America in 1909 to parasitize gypsy moth eggs, and it was highly effective. The wasp now has a sizeable population and will help control the moths near you as long as you steer clear of harmful chemicals.

Leaf Rollers

Leaf rollers are tricky to write about because the term refers to an entire family of moths (Tortricidae), which includes a host of different critters around the world. Leaf roller damage in Canada can be anything from harmless (though aesthetically unappealing) to devastating.

In Canada, besides the spruce budworm (see page 46), we're primarily afflicted by two species, which affect different trees but have similar life cycles and effects. The ash leaf cone roller (*Caloptilia fraxinella*) afflicts the Prairies, Ontario and Quebec, while the fruit tree leaf roller (*Archips argyrospila*) ranges across southern Canada and is most prevalent in British Columbia.

ASH LEAF CONE ROLLER

Ash leaf cone rollers are native to the eastern U.S. and have long been common throughout Ontario and Quebec. In 1999, they moved to Edmonton and have since touched all the Prairie Provinces. In the west they haven't become nearly the problem they are in the east, thanks to the prairie's devastatingly cold winters (yay for us...I think).

Watch the leaves of your ash trees for signs of leaf rollers.

IDENTIFICATION

The 1 cm wide grey moth is one of the first things gardeners will notice in spring. In summer, if you're sitting under your big, beautiful ash tree and getting a sunburn, you may want to look closely at the leaves. If they've been wrapped into tidy little samosa-shaped packages, you've got leaf rollers.

LIFE CYCLE

The adults lay eggs quickly in spring, and the resultant caterpillars mine leaves first and then, in June, move to another leaf to weave it together with fine silk thread and pupate.

DAMAGE

They afflict all types of ash, with the notable exception of mountain ash, which isn't related despite the name. The good news is that the damage is aesthetic and, while the reduction of photosynthesis can weaken the tree, it probably won't kill it.

CONTROL

Don't spray them with pesticides. The poison will damage the entire local food chain and won't penetrate the leaf to kill the critter inside. The damage is cosmetic.

There is a stingless parasitic wasp that has taken a liking to laying its eggs in the caterpillars and using their insides for baby food. As the leaf roller population grows, the wasp population will grow too, and will keep it in line.

Fruit tree leaf rollers are a major pest to apple and pear growers.

FRUIT TREE LEAF ROLLER

In the apple orchards of British Columbia and other Canadian regions, the fruit tree leaf roller makes the jump from cosmetic annoyance to crop-threatening pest. Although it can afflict ash, caragana, elm, maple and poplar species, it likes apple and other fruit trees the best.

IDENTIFICATION

Like other leaf roller damage, you can't miss it. The leaves will be rolled up like sloppily made cigars, and the fruit will appear scarred and ugly.

LIFE CYCLE

In spring, the larvae munch on emerging buds and leaves, weakening the tree from the start of the growing season. Unfortunately, they also enjoy the very small apples that are often forming as they feed, leaving the mature fruit scarred and commercially useless.

In June, they migrate via silken thread from the leaves they eat to the leaves in which they pupate, pulling one or several leaves around them for their sleepy metamorphosis. The emergent mottled brown and silver moths will lay 50–100 eggs in the crannies of branches to overwinter.

DAMAGE

While the damage to the tree itself is cosmetic, the damage to the pocket-book of people who are trying to sell damaged fruit is very real. Apples and pears will appear so scarred and warped that, though technically edible, they're commercially useless.

CONTROL

As with ash leaf cone rollers, spraying pesticides will do more harm than good. If you're a commercial grower and this leaf roller is threatening your harvest, consult a professional to help you deal with the problem.

Sod Webworms

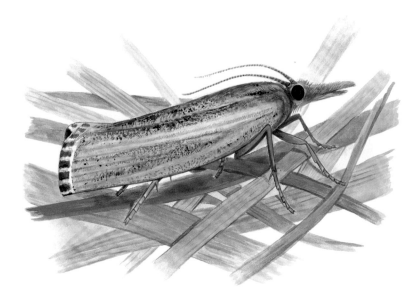

Does your lawn have brown patches that look like dog spots, only you don't have a dog? Sod webworms (Crambidae) are a common affliction in mature, exposed lawns (they don't like shade). Luckily, the best way to control sod webworms is also the best way to make your lawn healthier and greener.

Unlike most yard pests, webworms are American natives and weren't introduced from the "Old World." The adult moths do no damage, but the eggs they lay can wreak havoc on turf.

IDENTIFICATION

Got thatch? Webworms prefer mature, thinned-out lawns with plenty of cozy thatch to curl up in and dig their tunnels through. The more thatch you have, the happier they'll be.

Watch for brown spots from late June on. Webworms will often have several generations a year, so while the worms only chow down for a few weeks, repeated generations will make it look like they gorge for months.

Look for silken-lined tunnels in the thatch, typically occurring in the centre of brown spots. Other than webbing, their tiny green excrement pellets are a giveaway. You won't see the worms themselves, which are a light tan colour with brown spots, as they're nocturnal, unless you drive them out. The adult moths are mottled brown and are recognizable by their unique snout.

LIFE CYCLE

Webworm caterpillars overwinter deep in the soil and wake up in spring to start feeding. Most of the damage happens in mid to late summer as the larvae grow and move closer to the surface. In fall, they retreat deep into the soil again.

DAMAGE

The larvae will chew your grass blades off just above thatch level and pull them into their silken caves for edible bedding. The crown of the grass blade is typically below thatch level, so while they don't kill the blade, they will give it a nasty buzz-cut. However, if left untreated, the accumulating chomping stress may kill the lawn.

The damage will worsen during hot, dry summers. The scalped grass will still be alive, but chewed off whenever it tries to grow. Brown patches will eventually connect with each other and, if uncontrolled, there will ultimately be more brown than green.

CONTROL

Improving your lawn's overall health will go a long way in controlling webworm. Give it a spring and fall aeration followed by a good raking to pick up as much thatch as possible (elbow grease required).

After thatch detail, consider topdressing a spotty lawn with fresh seed. Mix some seed into top soil, sprinkle it across thin patches and water well with a fine spray. Healthy, thick lawns are less appealing to webworms.

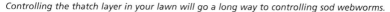

Controlling the thatch layer in your lawn will go a long way to controlling sod webworms.

Fertilize your lawn at least twice a year and, if the weather turns hot and dry, turn the hose on. The best way to water is weekly but deeply so that the water gets past the thatch and into the roots.

Webworm larvae make juicy snacks for a variety of predators, from ants, spiders and roving beetles to nematodes (which you can purchase at some large garden centres or online). A healthy backyard ecosystem will go a long way to control webworm numbers.

If you have only a few tell-tale brown patches, you can attempt to drive the pests out. Douse a browned area with insecticidal soap, either bought or homemade. Wait a few minutes, and they'll wriggle to the surface. Simply rake them up from there and be done with it.

BTK (*Bacillus thuringiensis* var. *kurstaki*) is a naturally occuring bacteria that loves nothing better than killing sod webworms. It's sold commercially and is a safe alternative to harsh chemicals such as Sevin.

Chemicals are like drugs; once you start using them, it's very hard to stop. They will wipe out your entire ecosystem, so when the next generation of webworms hatches, there will be no predators, the damage will be much worse and the temptation to reach for the chemicals will be that much greater.

A patchy lawn may indicate sod webworms.

Tent Caterpillar

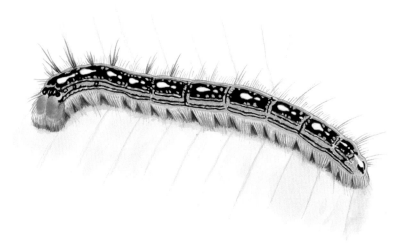

Anyone over a certain age remembers the "the tent caterpillar years," and most of us still have stories of how, when we were young, they'd sound like rain falling from trees and there would be so many on the road that cars would lose traction. Some years we may not see any, but the years we see a lot, we remember.

IDENTIFICATION

The term "tent" refers to a behaviour more than a specific species, so there's sometimes confusion over which tent caterpillar we're dealing with at a given time. Caterpillars that "tent" retreat into a silk tent during cool spring nights so their body temperature doesn't drop dangerously low.

Across the Prairies, the forest tent caterpillar (*Malacosoma disstria*) is the most ravenous. Mature caterpillars are about 5 cm long with blue stripes on their sides and white, keyhole-shaped marks on their back.

Tent caterpillars prefer the foliage of trembling aspen.

LIFE CYCLE

After hatching in spring, the caterpillars are mostly finished their pillaging by midsummer, when they retreat into cocoons to emerge as furry, brown moths. The moths will then lay the egg sacks that begin the cycle anew.

In our northern climate, the population goes up and down cyclically, with outbreaks occuring every 6 to 16 years, depending on conditions, and can last several years. While countless factors can make the outbreak either less or more severe (e.g., a hard frost while the eggs are hatching will really slow them down), make no mistake about it: the next outbreak is coming.

DAMAGE

Preferring trembling aspen, they usually keep to the forest, but on outbreak years, they'll squiggle their way onto city boulevards and into backyard gardens, devouring birch, ash, maple, fruit trees and cotoneaster.

While their defoliating ways don't kill trees directly, they will weaken a tree so that opportunistic and potentially dangerous pests and diseases can move in.

CONTROL

During the day, when they're dispersed and hungry, controlling tent caterpillars can seem impossible. A strong jet of water from a garden hose is an amusing way to send them flying, but the best way to really control their numbers is to target their egg sacks.

Female moths lay their eggs in brown, oval-shaped rings on aspen and other favourite trees, where they're clearly visible from leaf drop in fall through to their spring hatch. Scrape off whatever egg masses you can reach with a dull blade.

Rather than throw the egg masses away, leave them in a place where they won't be covered in snow, but where emerging caterpillars in spring won't have quick access to food. The egg masses are often home to beneficial parasitic wasps, so while the caterpillars will starve, the wasps will emerge to help control the populations that emerge from the egg sacks that were too high to reach.

For any caterpillars that manage to eat their fill, the next best opportunitity for control is to destroy their cocoons. The cocoons are grape-sized, white, silken clumps nestled on house siding, on trees and pretty much anywhere there's a small alcove. Destroying these will prevent the moths from emerging and, by extension, will prevent the eggs from being laid for the next season.

The damage is usually only aesthetic, but if you must resort to spraying, there are several environmentally friendly options available.

Look for egg masses on your trembling aspen and other trees in fall after leaf drop.

Spruce Budworm and Spruce Sawfly

If you notice damage to your spruce trees, one of two pests may be to blame.

If the tops of your spruce trees are brown and bare, chances are a hungry critter is munching on the succulent new growth. Lucky us, we have two different pests that could be ravaging our conifers. They aren't related to each other, and there are subtle differences to the damage they do, but you can kill them both the same way.

SPRUCE BUDWORM

The spruce budworm has been a destructive fixture in North American forests for hundreds of years. A pervasive threat to commercial forestry, it has become a significant pest in backyards across the continent.

The Prairies are afflicted with the eastern spruce budworm (*Choristoneura fumiferana*), while its western cousin (*C. occidentalis*) plagues the Rockies to the Pacific. Both species favour black spruce, white spruce and balsam fir, though in yards it goes for the spruce first.

BUDWORM IDENTIFICATION

If your trees are afflicted, the webbing of the budworms' silken shelters will be visible by the end of May. Look for it around needles and shoots, especially in tree tops. If you have mature conifers, you may need binoculars to check.

Western spruce budworm

The caterpillars appear in late May to early June and are a tan brown with markings across their bodies. They like new growth the best, so look for browning buds in summer. The adult moths that emerge in late summer are a dark, mottled brown.

If you have a keen eye, you can spot budworm eggs on the undersides of needles in fall; they look like drops of spittle. These can easily be washed off with a hose.

Budworm Life Cycle

After mating in summer, the female moths lay their eggs—usually about 150—on the undersides of spruce needles. The emerging larvae spin silken cocoons and huddle in the tree's nooks and crannies, dreaming of the darling buds of May.

When they emerge in spring, the caterpillars gorge on year-old needles, fresh buds and tender new growth first and only grudgingly eat mature needles once the new growth is gone. They feast for 40 days before pupating.

Budworm Damage

The combination of voracious appetite, duration of the caterpillar stage and preferred food makes budworms very destructive. Budworm damage is instantly recognizable in forests as the vast brown swaths of dead spruce trees.

Budworm Control

Getting budworms off small conifers is a straight-forward process of either blasting the eggs/webbing/caterpillars off with a high-pressure hose or, as a last resort, applying a pyrethrum-based spray.

With infestations on mature trees, height quickly complicates things. They usually afflict the tree tops, so you will need to get the high-pressure spray up high. If all else fails, pyrethrum or even malathion will work on the caterpillars, but both are contact sprays and need to touch the critter to kill it. Do not spray on windy days.

The best prevention, besides keeping neighbouring trees clean, is keeping your spruce trees healthy enough to fight off any budworm infestations. Plant them in the best spots possible and fertilize them regularly.

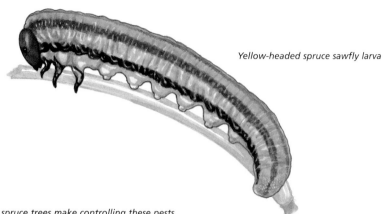

Yellow-headed spruce sawfly larva

Tall spruce trees make controlling these pests a challenge.

SPRUCE SAWFLY

We often confuse the yellow-headed spruce sawfly (*Pikonema alaskensis*) with spruce budworm, but although both pests attack new spring growth, they have more differences than similarities. While budworms are moths that overwinter enmeshed in webbing in the trees, sawflies are stingless wasps that overwinter in the ground.

Like budworms, sawfly damage becomes evident in late May or early June. They tend to attack treetops, but while budworms leave some remnants of needles behind, sawflies strip everything away. While budworms will strike every tree in a group, sawflies will strike a few but rarely all.

Sawfly worms are distinctive with a light green body and yellowish brown head. Treatment for them is the same as for budworms, only you won't be able to attack the eggs or webbing because sawflies overwinter underground.

Chemical treatment (pyrethrum or malathion) is most effective when sawflies are just starting to feed. This tends to be about 10 days after the brown bud caps have fallen off the tree.

Leaf Miners

Birch leaf miner

Have you noticed squiggly brown lines on your birch, lilac and vegetable leaves in summer? If so, you're not alone. Leaf miners are a common issue across Canada and afflict many types of plants. Rather than referring to a specific insect, "leaf miner" is a behavioural term that encompasses dozens of critters worldwide.

The good news about leaf miners is that their damage is cosmetic and, unless the plant is already very stressed, there's no lasting harm. The bad news is that, if you want to eradicate them completely, it's a long and somewhat tedious road.

IDENTIFICATION

A host of nasty critters "mine" leaves. Most of them are moths, but a few flies and sawflies do it, too. No matter what the specific insect is, it is always the larvae that do the damage.

Damage from a birch leaf miner

You can sometimes see the tiny larvae inside the leaf, but usually all you'll see are the winding, brown tunnels and the black "frass" they leave behind as they eat their way through your plants' leaves. By the time you see the damage, it is usually too late to stop it for that season.

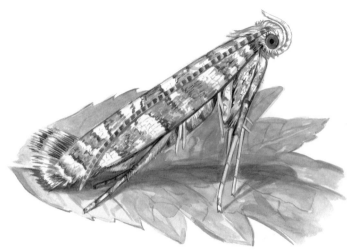

Lilac leaf miner

LIFE CYCLE

The insect, whether a fly, moth or beetle, lays its eggs on, or in, leaves, so that the hatching larvae have a buffet of tasty chlorophyll waiting for them.

The larvae leave tiny brown trails, resembling meandering rivers, as they chomp through the soft leafy insides. Eventually, the trails merge with each other, and meandering lines become brown splotches. This usually represents the end of the larval stage and is the worst the damage will get.

Damage from a lilac leaf miner

Although leaf mining insects have about three life cycles a year, we usually don't notice the damage until the third and last one in midsummer.

The insects overwinter by burrowing into the ground near the host plant, emerging the next spring to start all over again.

DAMAGE

Although losing leaf tissue weakens plants by reducing their ability to photo-synthesize, the damage is seldom fatal. Keep them well-watered and fertilized, and they will soldier on. If your plants are badly affected, keep your eye out for other, hopefully more easily treatable, opportunistic pests or diseases that can strike a weakened host.

Birch trees are commonly afflicted with the birch leaf miner (*Fenusa pusilla*) and the amber-marked birch leaf miner (*Profenusa thomsoni*), both of which were introduced from Europe about 100 years ago. These are tough critters to get rid of, but the damage is usually cosmetic; expect the population to go up and down over the years.

Vegetable leaf miner

CONTROL

Apply horticultural oil as soon as the buds start unfurling in early spring, and apply weekly until June. Although you'll need to persist for several years, this is usually an effective treatment.

For beets, tomatoes and other smaller plants, I suggest opting for diversionary tactics over full-out conflict. Planting nasturtiums and columbine near the affected area will lure leaf miners from your prized vegetable and tomato plants.

There are no effective sprays for leaf miners while they're inside the leaves. All pesticides on the market today are contact killers and won't touch the larvae. If you time it perfectly, you may be able to hit the critters as they're emerging from the ground in spring.

Several species of parasitic wasp feast on emerging leaf mining critters and can help keep populations in check. Try to avoid the use of pesticides in order to protect any populations of predators you may have.

Leaf miner damage on a honeysuckle

Dew Worm

Although earthworms are excellent to have in the yard for their ability to aerate the soil, dew worms (*Lumbricus terrestris*), a particularly large form, are far too much of a good thing. Active at night and early morning, when the lawn is wet, they can churn a healthy lawn into a chaotic jumble of mounds and deep tunnels.

IDENTIFICATION

If you've never seen a dew worm, do I have a mental picture to paint for you. Imagine the cute little red wrigglers in your garden, only reaching nightmarish proportions. Dew worms are worms the size of snakes, with their slimy, gummy bodies ranging from 10 to 30 cm long.

When you walk across your lawn in bare feet, does the ground under the grass feel like a miniature Western Front, complete with shell holes and tossed mounds of earth? If so, you've probably got dew worms.

Dew worms thrive in old lawns, especially those with decades of accumulated thatch. Besides age, dew worms love shaded, sheltered, well-watered lawns.

LIFE CYCLE

Earthworms, of which the dew worm is the world's largest type, are hermaphrodites. Every worm has both male and female organs, which makes their mating experience a little out of the ordinary.

Worms still need two to tango, but the process is unique. They exchange sperm with each other by rubbing their bodies together, and then cocoons form on their bodies, each one containing about five baby worms.

Their tendency to reproduce when they're only a few months old, combined with their ability to live for years in ideal conditions, means that worm populations can boom very quickly. When worms die, their bodies simply dissolve into the earth as fertilizer.

DAMAGE

The mounds in your lawn, which can be irritating to both walk on and mow over, are castings pushed up from the worms' burrowing. Although the worms won't kill your lawn, severe infestations will make it almost impossible to walk and play on; dew worms will effectively evict you from your own yard.

CONTROL

Underneath the casting mounds, dew worms' extensive tunnels can dip several metres below the surface. This inaccessibility, combined with their bafflingly creepy size, makes them very hard to kill.

If you won't be satisfied until every one of them is dead, you have a frustrating road ahead. My recommendation is to focus on controlling their numbers, and the damage they do, by keeping them underground. They're also highly mobile and tend to populate blocks of houses at a time, often in older neighbourhoods, so if you use chemicals to eliminate them, you are really only creating a vacuum and encouraging others to move in. If you use cultural, preventative controls, you will make your yard less appealing in the long run.

Aerate your lawn in spring and fall, and rake it out well to reduce the thatch layer and make the surface less appealing to worms. There are many companies offering the service, or you can simply rent an aerator.

Don't water your lawn in the evening if a cool night is expected, and only water when needed. In a normal rainfall year, a healthy lawn actually needs little or no

Dew worms love shaded, sheltered, well-watered lawns.

supplemental watering. If your lawn is patchy, top dress with fresh grass seed or, if it's very unhealthy, consider tearing it up and starting again with fresh turf.

As with all soft-bellied creatures, dew worms hate crawling over abrasive surfaces (which is why they love wet grass at night). Sprinkle a generous layer of sharp sand over the affected lawn. While nonlethal, it will make dew worms think twice about venturing to the surface.

If you insist on using chemicals, the active ingredient carbaryl is approved for use against dew worms.

Most products containing carbaryl have been banned, and the last product with it is called Sevin. Sevin will reduce your dew worm numbers, but even it won't eliminate them. It is definitely not safe for animals, children or pregnant women.

Please regard Sevin as an absolute last resort (I don't recommend it as any resort). You will effectively wipe out the entire ecosystem of beneficial bugs, fungi and bacteria that have built up in your yard. If you use Sevin, expect a host of other pests such as aphids to spring up because you've just nuked all their predators.

Aerate and dethatch your lawn every spring and fall to make the surface less appealing for dew worms.

Slugs

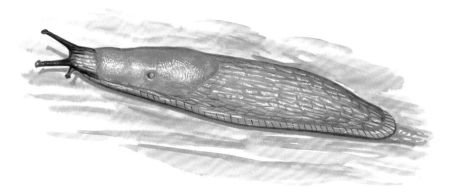

If Canada had a "Most Disgusting" award, slugs would win it every year. These gross gastropods run rampant during wet summers, but fortunately, they aren't hard to get rid of.

IDENTIFICATION

Technically speaking, slugs are any gastropod molluscs without a shell. During wet summers, slugs set up camp in the perpetually damp, decaying organic matter that forms a slimy mulch in your flower beds.

Slugs aren't subtle. Overnight, wide-spread holes will appear on your leaves, usually accompanied by dried, glisten-ing slime trails and sometimes clumps of tiny black droppings.

If you suspect you have slugs, dig through the organic matter and com-post at the base of the affected plants. You can also sneak up on them first thing in the morning, before the dew evaporates.

To be sure that slugs are the culprit, check for caterpillars because their damage is similar. Unlike slugs, you'll easily spot caterpillars. By the time you look for slugs, they are usually safely tucked into their rotting daybeds.

Slugs love to chew through hosta leaves.

LIFE CYCLE

As far as garden pests go, slugs are big creatures with a gluttonous appetite. They're hermaphrodites, so they basically reproduce at will, laying clutches of eggs that resemble rotten caviar.

Slugs live in the wettest, shadiest and rottenest piles of leaves, weeds and muck that collect in the dark corners of your perennial and flower beds. They surround and protect themselves in the most cluttered spots they can find, and that's where they spend their winters.

DAMAGE

A slug infestation can turn a bountiful lettuce crop or your prize hosta collection into a smattering of tattered stumps in a matter of days. With slug infestations, it's usually them or your plants. Slugs can bring long-term harm, as plants without leaves will have a hard time taking in the nutrients they need to survive our menacing winters.

CONTROL

Fortunately, there are many ways to control slugs. Although the first one that comes to mind may be chemical slug bait, I encourage you to consider this a last resort. You'll kill the slugs, but you will probably kill many beneficial critters, too.

The best way to keep them from moving in in the first place is to clean up the neighbourhood. Every fall, clear out leaves, sticks and assorted junk from your flower beds. Remember that slugs will also happily hibernate under non-organics such as hoses and forgotten tools, so clean those up too.

Try to keep the garden clear of debris throughout the season as well. This includes old, decaying mulch that may need to be replaced. While you're at it, carefully till and fluff the first few inches of soil; slugs often crawl into it to sleep.

Of the many ways to get rid of them, I've found that beer works the best. Submerge a margarine lid in the garden so the lip of the lid is flush with the soil. Fill it with cheap beer, leaving the good stuff for yourself. They will crawl into it and drown. To get them all, it will need to be replaced daily; they like stale beer as much as we do.

You can also go out at night and kill them manually. They are easy to spot with a flashlight, and a sprinkle of salt (and a strong stomach) is all you need.

Some people swear by copper, saying it shocks them as they smear across. You can buy special copper wire or bands, or just sprinkle pennies around your plants.

You can also lay down a perimeter of diatomaceous earth around the plants. It's comprised of very small, very sharp silica crystals that act like razor wire on gelatinous slug bellies. Wear a mask when you apply it.

Other home remedies include dried eggshells (they cut slugs), wood ashes (too alkaline for slugs) and used coffee grounds (slugs don't like caffeine). The reviews I've heard about these methods range from mixed to awful.

Lastly, the best control is the most natural. If you have frogs, ground beetles, salamanders or (best of all) garter snakes in your yard, encourage them to stay because they happily eat your problem away; grow a diversity of plants, don't use chemicals, and create hiding places for them.

This woodland garden is prime slug habitat.

Root Maggots

Onion and cabbage lovers beware! While your favourite vegetables grow, there might be a greedy pest doing dastardly deeds under the surface, devouring your pungent bulbs and your sweet and spicy crucifer tubers. Onion maggots (*Delia antiqua*) can destroy fields worth of onions, garlic, shallots, leeks and chives; red onions and Japanese bunching onions have shown some resistance. Cabbage root maggots (*D. radicum*) will tunnel into your radishes, turnips and rutabagas.

IDENTIFICATION

Root maggots are hard to get rid of once they're established, so it's vital to keep a keen eye out. They're most virulent during wet, cool growing seasons.

The adult flies resemble common house flies with a few differences. They're ash grey with a humped back, overlapping wings and brown eyes. The maggots are up to 1 cm long, white, tend to cluster together and are rarely seen.

The first sign of an infected plant will be wilted, yellowing leaves.

Onion maggots can destroy your entire row of onions (above), but red onions (below) have shown some resistance.

LIFE CYCLE

Root maggots overwinter in the soil as stubby brown pupae, usually very close to where they feasted on host crops the year before. You can often find emerging flies wooing each other around dandelions' first spring flush (how romantic). They can mate less than a week after becoming mature flies at six days old, and then begin laying their eggs at the base of fresh plants (often the ones that you've just planted) three to four days later. A female fly can lay up to 200 eggs in several odious broods in its month-long lifespan.

Root maggot eggs hatch in a week, and the critters unleashed are ravenous. They need to grow to 1 cm long before they can burrow into the soil and pupate, and they will feed on a bulb or tuber for over two weeks to get what they need. If the plant dies before they are done, they simply move on to the next one.

Pupation during summer lasts two to three weeks, and the second-generation adults emerge to lay eggs again. This second generation overwinters in the pupal stage, beginning the cycle anew.

DAMAGE

You'll be able to see the leaf damage by mid-June. Upon pulling the plant out, smaller bulbs and tubers might be hollowed out completely. First-generation damage often results in the death of your seedlings.

Second-generation damage to large bulbs and tubers later in summer may be harder to see; look for small brown entrance holes. If you miss the damage and harvest the vegetables anyway, either they will rot in storage or the damage will be shockingly apparent as you prepare your evening meal.

CONTROL

Although root maggots can have two or three generations per year (depending on the climate), the first one, which hits the bulbs and sets you've just planted, is often the most destructive. Delaying planting for a week or two might buy you enough time to avoid that round.

Protect your seedlings with row covers when you plant them to keep the egg-laying flies off. Or hang yellow sticky traps around emerging plant shoots in spring in the hopes of catching the flies as they zero in. I've also heard that sprinkling the soil around the seedlings with a generous helping of cayenne pepper will help deter them.

Work the soil in fall, after an infestation. The pupae will be hoping to overwinter in peace and won't appreciate a hearty pitchfork toss. If the area is small and the infestation great, consider replacing the top layer of soil entirely. No matter what, make sure to remove all infected plants and destroy them as soon as possible. Do not throw them in the compost bin.

Rotate your crops. If the flies can't find your young plants, they can't lay their eggs. Or if all else fails, grow your pungent favourites in a container. Pick a spot some distance away from the previous years' infestation and use clean, fresh soil. Make sure the container is at least 30 cm deep so the roots have room to develop properly.

Rutabagas are at risk from cabbage root maggots.

Colorado Potato Beetle

The Colorado potato beetle (*Leptinotarsa decemlineata*), alias potato bug, alias ten-striped spearman, is the most destructive potato crop pest in Canada. After entering Canada via Ontario in 1870, it has become a ravenous and economically destructive pest across the Maritimes, hitting New Brunswick and P.E.I. especially hard. It's a nuisance across the Prairies, though not as pervasive.

Unlike many garden pests, which originated in Europe and Asia and invaded across the oceans, the Colorado potato beetle came from…wait for it…Colorado. From there, it has spread almost worldwide and has shown a remarkable ability to become immune to even the harshest insecticides (it even developed a defence against DDT in 1950s).

IDENTIFICATION

You'll know one when you see one. About 1 cm long, adults sport bold, brown stripes down the length of their golden back. If they weren't so destructive, we'd say they were downright beautiful.

The larvae are striking, with a dark red body, black head and jaws big enough to rip your potato leaves to smithereens.

LIFE CYCLE

Potato beetle adults overwinter in the soil around their victimized potato plants. In spring, they emerge to seek host plants, feed for a few days, and lay up to 500 yellow or orange eggs in early July. The eggs hatch in less than a week, and the resultant larvae spend a few weeks after that chowing down.

Colorado potato beetle larva (above & below)

Although they can produce multiple generations per year in warmer countries, in Canada, the prohibitive climate typically limits them to one generation; being cold has its advantages.

DAMAGE

These beetles are ravenous and can strip potato fields bare with alarming speed. The larvae can chomp up to 40 cm² of foliage per day, per critter. Multiply that by each female laying up to 500 eggs, and the math proves terrifying. Adults clock in at 10 cm² of foliage per day.

Plants without leaves can neither photosynthesize nor produce big spuds. Eventually, weakened plants will simply die from stress.

CONTROL

Make sure to rotate your potato crop every year so that the emerging beetles have a tougher time finding hosts in spring. Laying down an organic mulch, such as straw, can also confuse emerging beetles looking to lay eggs long

Rotate your potato crop every year to protect it from this nasty beetle.

enough to keep them wandering until starvation.

Potato beetles have a host of predators, and some of them have even been harnessed for commercial use. From ground beetles, parasitic wasps and nematodes to a pathogenic fungus, sometimes your garden army will mobilize itself to keep invaders under control. Curbing your chemical use in all areas, and fostering a healthy ecosystem, will go a long way to supporting the predators.

Colorado potato beetles are famous for their ability to withstand and evolve against insecticides, so applying chemicals will do more harm than good. After all, this is an insect that lives exclusively on leaves from the Nightshade family, which are toxic to almost everything else.

A World Player

Not many bugs have been accused of being covert Cold War agents. Potato beetle history reads like a spy novel. In the early 1950s, the Warsaw Pact accused the CIA of air-dropping potato beetles across East Germany to destroy their crops.

After fields that American planes flew over were suddenly afflicted, a propaganda campaign, complete with posters, sprang up. Children across East Germany were sent out after school to collect and destroy the beetles, which were labelled amikafer, or "Yankee beetle." The allegation was never proven one way or the other.

In the 2014 Ukrainian conflict, pro-Russian rebels sported colours of St. George to show their allegiance to Moscow. The black and gold armbands have earned them the derogatory nickname kolorady, meaning "Colorado beetle" in both Ukrainian and Russian.

Crucifer Flea Beetle

Of the several flea beetle species afflicting Canada, the crucifer flea beetle (*Phyllotreta cruciferae*) is the most damaging. Native to Eurasia, it invaded North America in the 1920s.

Infamous for its love of young canola, it costs affected North American farmers about 10 percent of their annual canola crop, a loss worth about $300 million. When it can't get canola, its attention turns to the nearest gardens sporting broccoli, cabbage and other edibles and ornamentals.

The related spinach flea beetle is another vegetable garden pest.

IDENTIFICATION

The aptly named flea beetle is tiny enough to be hard to find (2–3 mm long), though its tendency to hop when disturbed often gives it away. While the crucifer has a black shell, other less common species are striped or metallic grey.

The larvae are tiny, grub-like creatures with an off-white body and a brown head.

If it looks like your garden has hail damage but there hasn't been a storm, you've probably got flea beetles. They make leaves look like they've been used for BB gun target practice.

LIFE CYCLE

Like the potato beetle, flea beetle adults overwinter in soil near their last victims and emerge in spring to find new ones. They develop a shameless bout of spring fever and mate repeatedly throughout May. Females lay up to 100 eggs in multiple batches, which hatch in two weeks and unleash the hunger anew.

DAMAGE

Crucifer flea beetles are a serious agricultural prairie pest. They typically overwinter in or near canola fields and are a worse pest during springs when the canola crop starts early. When spring is wet enough to significantly delay the crop, the flea beetles tend to terrorize neighbouring gardens instead.

The adults do most of the damage as they feed and mate throughout May. When they can't get canola, any cruciferous plant will do. This group includes broccoli, cabbage, turnips, Brussels sprouts, alyssum and even nasturtiums. Damage is most severe when the plant is young (even a seedling) and at its most vulnerable.

CONTROL

Given that they overwinter in leaf litter, weeds and other garden detritus, a thorough fall clean-up will go a long way in controlling them.

Get in the habit of rotating your edible crops (all of them), as this deters many overwintering pests, which tend to be picky about the plants they eat. They will emerge in spring only to find their favourite food gone.

Flea beetles have ample natural predators, but unfortunately, the beetles tend to emerge in such massive numbers in spring that the predators are overwhelmed. Planting marigolds and fennel will encourage more beneficial predators (such as non-stinging parasitic wasps) to call your yard home.

Flea beetles loathe thyme, catnip and mint, making companion planting an effective passive control method. If you're afflicted, sprinkle these plants amongst your cruciferous crop to help deter egg-laying flea beetle females.

All young crucifer vegetables, including turnips (above) and cabbages (below) are a target for crucifer flea beetles.

Cabbage Worm

If you love to munch on broccoli, cabbage, cauliflower and the crisp, broad leaves of kale, you're not alone. While we wait patiently for our crops to mature, there's a hungry worm waiting, too.

The cabbage worm (*Pieris rapae*), alias cabbage white butterfly, alias imported cabbage worm, was introduced to Quebec in the late 19th century. It quickly spread across Canada and today afflicts Brassicas from coast to coast.

IDENTIFICATION

Cabbage worms are pudgy green eating machines. They're about 2.5 cm long, velvety green and covered in tiny hairs.

You'll probably see the worms' handiwork before you find the worms. Look for large, irregular holes across your leafy greens. Cabbage worms are voracious, and two or three worms per plant can make it look like there's an invading army.

The adult butterflies are a distinctive light grey with two darker grey spots on each wing. You'll see them flitting about the garden at different times of the year; they can have four complete life cycles in one growing season.

LIFE CYCLE

The worms spend the winter as pupae, tucked under the soil. They emerge in spring as adult butterflies to lay their eggs on the undersides of plant leaves.

The eggs take only about six days to hatch, and the emergent caterpillars eat for a few weeks before pupating and starting the whole process again. They'll start out slow in spring, foraging for what they can get, but as your crops get bigger in summer, so will the worm population.

DAMAGE

Besides causing unsightly holes in cabbage leaves, the worms can eventually kill the plants by chewing away their ability to photosynthesize. Leaves with a few holes in them are perfectly edible, however, and in some European restaurants are even sought after as evidence of a truly organic salad.

CONTROL

The good news is that cabbage worms are delicious—if you're a predator—and a host of spiders, wasps and beetles feast on them. Unfortunately, the insects' speed of reproduction usually makes other control measures necessary. However, if you rely on frequent chemical use to solve other issues in the yard, you may find that the cabbage worm population builds up exponentially because many predators have been wiped out.

If the marauding butterflies can't land on your veggies to lay eggs, the hatching larvae can't eat everything. Placing floating row covers over susceptible vegetables when you plant is an excellent physical deterrent. Dropping large (5 cm) pieces of egg shell around the affected area, in theory, will resemble the butterflies and

The large, irregular holes in your cabbage leaves may be the work of hungry cabbage worms.

If your kale is afflicted, there are several control options available.

act as a scarecrow for other moths and butterflies looking to lay eggs, for a different type of physical deterrent.

The size of the worms, and the fact that there typically aren't more than a few per plant, makes them easy to pick off manually (unless you have a yard full of kale). Follow the holes and green droppings; they're typically on the leafy undersides or near the centre. It may be tricky to pick them off Brassicas that form heads, like cabbages, as they often burrow into the forming centre to munch on tender leaves.

If you like chasing butterflies, or have kids who do, you can always grab a net and chase them around the yard. It sounds taxing, but every butterfly removed is about 200 eggs taken out of circulation, not to mention the generation after that.

If preventative measures fail, there are organic options that don't require spraying chemicals on your future salad. BTK (*Bacillus thuringiensis* var. *kurstaki*) is a naturally occurring bacteria that loves killing worms and caterpillars. It has no known toxic effect on humans, birds, pets, bees or other beneficial insects. Hot sun will kill the bacteria after a few days, so you'll need to reapply following heat waves.

Cutworms

Cutworms are a group of caterpillars (Noctuidae) that can lay waste to large swaths of gardens. Commercially, cutworms are a major threat to Prairie farmers, with the potential to ruin crops of canola, corn and more. In the vegetable garden, they'll chew through corn, tomatoes, peppers, cabbages, beets, peas, beans, squash, leafy greens and more.

There are dozens of species of cutworms. They all do the same damage, however, and they all curl up into a protective ball when discovered.

IDENTIFICATION

Cutworms can be brown, grey, pink, green or black, and spotted or striped. They burrow amongst leaf litter or in the soil during the day and emerge to feast on your plants at night.

LIFE CYCLE

After their spring feeding frenzy, the caterpillars will retreat into the soil to pupate into brown, mottled moths that emerge in summer to flit harmlessly around the garden. In fall, they'll lay 1000 or more eggs in leaf litter and soil. The eggs hatch shortly after, and the larvae feed briefly and then overwinter in the soil. In the first warmth of spring, the larvae emerge, and the voracious festivities begin anew.

DAMAGE

Cutworms are aptly named for the kind of damage they do. They're nocturnal, and they wiggle out of their dirty daytime hiding places at night to chew through the first plant bits they find, which is usually the base of the plant. Once the plant topples, the caterpillar typically moves on to another plant in a very wasteful feeding pattern that usually leads to widespread damage.

Cutworms prefer tender seedlings to established plants, often toppling whole rows of delicious edibles in a single night, leaving only decapitated plants and a baffled gardener by morning.

CONTROL

If you have cutworms one year, then you can bet your bottom dollar the larvae are as impatient for the following spring as you are. The first way to control them is to deny them their initial feeding frenzy. Delay spring planting for a week or so after you would normally plant (it will be painful, I know) to starve them out.

If you have only a few plants, you can wrap cardboard collars around the stems. For a whole garden, however, it would be faster and easier to get on your hands and knees and pick the cutworms out of the soil individually.

They hide during the day, so if you want to do the ol' find-and-squish, you need to either go out at dusk, dawn, or ideally during the night. If nighttime hunting is not your thing, during the day, you can drench the suspect area with a mixture of 4 L hot water to 5 mL dish soap and wait for the cutworms to wriggle, gasping, to the surface.

As you would for slugs, you can sprinkle diatomaceous earth or crushed eggshells around your plants to curtail their movements. I've heard, though only anecdotally, that coffee grounds also work.

Young vegetable seedlings, including tomatoes (left), corn (above left), peas (above right), beets (below left) and kohlrabi (below right) are all susceptible to cutworms.

Mow your lawn regularly and keep leaf litter and weeds to a minimum, especially in late summer and fall. Just before freeze up, cultivate your soil in the afflicted area.

Flipping the first few inches will expose many of the overwintering pests and, usually, hungry birds and other predators will make short, delicious work of them.

Mosquitoes

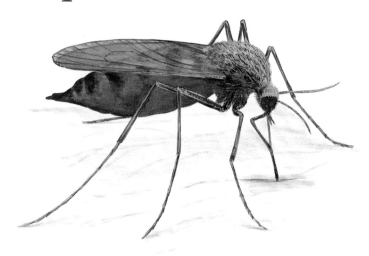

They are the bain of backyard barbecues everywhere and something that we've had to learn to live with. Mosquitoes are an inevitable part of summer, but just because we can't escape them doesn't mean we can't reduce their populations.

As persistent as they are, mosquitoes need some very specific things in order to hatch and bite. The easiest way to control them is to control the conditions that allow them to spawn. After that, you can use their sense of smell against them to protect yourself.

MOSQUITO 101

Of the thousands of species of mosquito (Culicidae) around the world, over 100 of them make Canada their home. Their biggest vulnerability is that they need water to hatch and thrive, which is why they are most prevalent (and annoying) in wet summers.

Their life cycle is simple. The eggs, often laid in fall to lie dormant all winter, hatch when they are exposed to standing water. The larvae, called "wrigglers," live and grow in the water for a week or so before molting into pupae. The pupae, or "tumblers," are strange, alien-looking creatures that stay in that form for only a few days before emerging as hungry adults.

Male mosquitoes don't bite, but their ability to fertilize eggs makes them fair game for smacking. Males have a feathery proboscis as opposed to the smooth needle that females have. They feed on flower nectar and typically live for only a week.

Female mosquitoes do the biting as they look for a blood-meal in order to get protein with which to develop their eggs. That's where we come in as targets of opportunity.

CONTROL 1: DRAIN STANDING WATER

Mosquitoes' quick life cycle (some species can sprint from egg to annoying in less than a week) means that you always have to monitor your yard for changing conditions. Luckily, they need a very specific, and often controllable, element: standing water.

Eliminating standing water is easy, organic and very effective. Do a yard audit by walking every inch of it. Make sure to get off your usual pathways and stick your nose behind bushes and around sheds. Look for old wheelbarrows, forgotten birdbaths, half-buried tarps and anything else that can accumulate rain water. Drain every splash of standing water possible. Mosquitoes tend to stay close to where they pupated, so, barring strong scattering winds, if your yard is free of standing water, you should have far fewer mosquitoes.

Next, put a weekly reminder in your calendar to drain and replace the water in your birdbaths, wading pools and any other deliberate standing water zones. It needs to be done often to keep mosquitoes' quick life cycle at bay. You don't need to worry about running or treated water.

Consider sprinkling larvacide on the surface of standing water that you can't drain. This is biological warfare that, while safe for humans and pets, unleashes an army of hungry predators to attack the wriggling mosquito larvae.

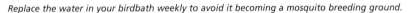

Replace the water in your birdbath weekly to avoid it becoming a mosquito breeding ground.

A neatly trimmed yard means a less hospitable habitat for mosquitoes.

Not many people associate eaves-troughs with mosquitoes, but partly clogged troughs (if you have big trees in the yard, that probably means you) create pockets of ideal skeeter habitat. Clean those gutters in spring and check them periodically.

CONTROL 2: INHOSPITABLE HABITAT

We've all walked through damp summer grass and seen the biting hoards buzz upward en masse like we're in a bad horror movie. Although it's a myth that mosquitoes breed in tall grass, they certainly like to rest there. Dew provides a refreshing drink, and there's a built-in alarm system (i.e., your feet) to tell them when dinner is near.

Mosquitoes have no love of sunlight. They're most active at dawn and dusk and dwell in cool, still shade. Keeping your shrubs and trees neatly pruned and your lawn trimmed will reduce hiding and resting places (not to mention make your yard look better).

CONTROL 3: MOSQUITO REPELLING PLANTS

Inevitably, some mosquitoes will find their way into your yard no matter how much water you drain and how neatly trimmed your yard is. Let's face it, to them you're a buffet and a necessary part of completing their life cycle. But you can make them want to leave with a few well-placed plants.

When people think of mosquito-repelling plants, they automatically think of citronella geraniums. While they are often seen as the most effective, there are other plants that work as well.

Mosquitoes hate marigolds. They avoid the pretty yellow flowers for the same reason that garden pests do; they contain the natural pesticide pyrethrum.

Ageratum is another annual that mosquitoes avoid, thanks to the smelly coumarin the flowers emit; coumarin is a chemical widely used in mosquito repellents.

The plant with the most scientific evidence of repelling effectiveness is catnip, a perennial herb closely related to mint. While our kitties roll around in it, bloodsuckers loathe it. Some people even go so far as to rub catnip leaves on their exposed skin. Although this will probably keep the mosquitoes away, cat owners be warned that you'll smell like a giant sprig of catnip to Fluffy.

An important caveat: plants are not DEET, and they don't work with the industrial strength of synthesized chemicals. They are a passive repellent that, while not 100 percent effective, don't have a foul odour or make you wonder about rubbing yet another chemical on your body.

Catnip is a proven mosquito repellent.

Marigolds will keep mosquitoes away.

CONTROL 4: DEFENCE!

One of the best parts of summer is hanging out, usually around twilight, in the yard with a cold beer or three. Don't let rampaging skeeters chase you inside. It's surprisingly easy to establish a defensive perimeter around your favourite hang-out spot.

Know your enemy. Mosquitoes are weak flyers and shun wind. Keeping them off your patio can be as simple as setting up a few fans to circulate the air around you.

For more traditional methods, there are a lot of citronella candles, zappers, smokers, etc., to choose from. There's even a phone app that emits a supersonic pulse to keep them at bay, but don't use it if you want your pets to come around.

Ants

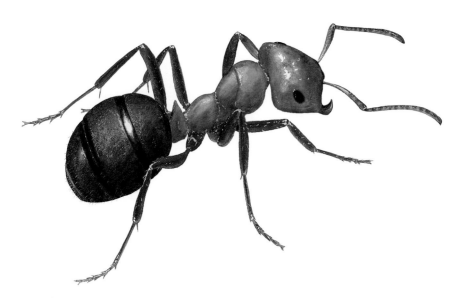

Field ant

Before we talk about killing our ants (Formicidae), let's take a moment to be thankful for living in one of the least ant-infested countries in the world. I've seen leaf-cutter ants strip shrubs bare before my eyes and almond-sized hunter-killer ants that will make a grown man's arm go numb for days. I've watched a swarm of army ants take 20 minutes to march past me and seen Cambodian villages built entirely on stilts sitting in water just to escape the swarms.

Even as we scatter, mow and boil these critters into submission, we should remember how incredible they are. An ant colony functions so seamlessly that it could almost be thought of as one organism rather than a group of individuals. Colonies can be millions strong and thousands of miles in breadth. Ant societies have complex class structures, keep slaves and herd other insects like cattle in order to serve the collective.

If you combine the weight of every ant in the world, it would be heavier than the combined weight of humanity. They're indigenous to every landmass except Antarctica and have been around since dinosaurs wondered how to get rid of them. To be blunt: ants aren't going anywhere.

ANT 101

Of the estimated 22,000 species of ants worldwide, Canada has a few hundred of them. Depending on where you are in the country, you're likely to have different encounters.

The ants in your yard are probably the common field ant (*Formica* spp.). Their long list of aliases includes wood ant, thatch ant and mound ant. These ants are either black, red or bicolored, and are generally harmless and may even be beneficial. Their tunnelling aerates the soil, and they often prey on nastier critters. If they're living in a little travelled area and not offending you, consider living and letting live.

If the ants are small and yellow or light brown, almost transparent, they're potentially invasive pharaoh ants (*Monomorium pharaonis*), and live-and-let-live goes out the window. Red harvester ants (*Pogonomyrmex barbatus*) have a painful bite and should also be taken care of right away, as should carpenter ants (*Camponotus* spp.), which are large, almost black and have a triangular head.

Ants love loose, sandy soil, which makes old lawns with depleted soils (typically rarely fertilized) favourite haunts. If your lawn is struggling, it's going to be less likely to grow aggressively enough to repel ants.

They often establish colonies in and around root systems to take advantage of the natural tunnels therein. This tendency can spell trouble for many plant species and is the reason ants are so prevalent in veggie gardens, where you've kindly tilled the soil for them. Ants also have a famous sweet tooth (well, sweet mandible). They will swarm into yards littered with tasty cherries, apples or other fruit.

If you want or need to evict them, you're not alone. Ant hills are unsightly and can spread quickly. Ants can also increase nearby populations of aphids, which they farm—and actually herd from plant to plant—for their tasty honeydew secretions. If the nest is near your home, eradicate it and check for cracks in your foundation that foragers might wander into.

Harvester ant

Carpenter ant

CONTROL 1: WATER

The earlier you catch them, the better. Field ants build conspicuous mounds in brightly lit locations. The more you kick, mow and soak them down, the more likely they will be to abandon the spot in frustration. Give the anthill a few good scuffs with your boots, and if you have a self-propelled lawnmower, give them some spinning wheels.

Perhaps the most effective control method is to soak them out. Ants hate excessive water, which makes them more prevalent during dry spells and in un-irrigated yards. The caveat to this is that while you may see them more during wet years, it's only because the wet earth has forced underground dwellers to the surface, like worms after a rain.

The business end of a hose is surprisingly effective, even against hard-to-reach, under-the-sidewalk and in-driveway-crack colonies. Soaking the hills while you water your flowers collapses their tunnels; keeping them mud pits makes new tunnelling impossible.

If the hose doesn't work, weaponize the water. Bring a large pot to a rolling boil. With a spade, dig deep into the hill and turn it over, ideally exposing the queen and nursery chambers. Scald the works with boiling water; they hate that.

CONTROL 2: HOME REMEDIES GALORE

While there are several chemical control options—with varying toxicity—I suggest them only as a last resort. There are a host of organic ways to either kill ants or convince them to move.

Here are just a few of the home remedies I've come across. A quick internet search will reveal dozens more, with vastly varying degrees of effectiveness or just plain kookiness.

A lot of people rave about how effective borax, an old-fashioned cleaning agent, is for controlling ants. Typically, it's mixed with icing sugar or jam to act as the proverbial poison apple. While borax contains boric acid (a common ant-killing ingredient), they aren't the same thing. Borax is naturally occurring and hasn't been found to be either carcinogenic or accumulating in the body. However, if you use it, wear gloves (it's a skin irritant) and a mask, and use it in moderation.

Sprinkling diatomaceous earth around the colony is an effective way to make them move. It won't kill them, but the sharp silica crystals are like razor wire to ants. Wear a mask while applying it. You can buy it at most large garden centres.

As for other things you can sprinkle on and around the infested area in order to clear the ants: cornmeal and icing sugar, baking powder and icing sugar, used coffee grounds, minty essential oils, cream of wheat, Equal (the sugar substitute), vinegar, cinnamon, cucumbers, mint leaves, table salt and chalk (the rumour is that ants can't walk across chalk).

Any cherries that fall to the ground will attract ants (above). Ants will actually herd aphids from plant to plant, collecting the aphids' honeydew secretions as they go (below).

Wasps

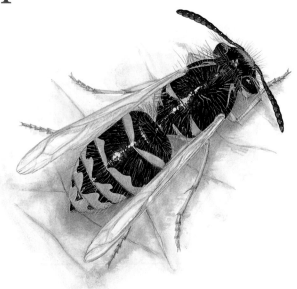

Yellow jacket

Few insects are as universally unwelcome as wasps. But before I tell you how to get rid of them, I'm going to tell you how fascinating they are.

Of the over 100,000 species of wasps worldwide, the majority are parasites that look nothing like their infamous cousins. Parasitic wasps are harmless to humans but barbarous to the insects they inject their eggs into. At the greenhouse, many of the best predatory insects we use for our biological pest control are microscopic wasps.

Only a small fraction of them (namely the social wasps in the Vespid family) will sting humans. These are the nasty, picnic-crashing wasps we all know and dread. Whether you're scooping ice cream or biting into a juicy burger, there's a good chance that wasps will show up.

WASP 101

When it comes to wasps, the most troublesome types are yellow jackets, hornets and paper wasps. They look much different than honey bees and bumble bees, being much thinner and sleeker, and are in fact not even related to them.

Yellow jackets (*Vespula* spp.) love sugar the most, so they're most likely to crash your pop- and pie-fuelled picnic. Their nests are typically underground, so lawnmowers in shorts beware. Yellow jacket is actually the common name of the entire genus of nasty Vespid wasps that we're most familiar with.

Bald-faced hornet (above); paper wasp (below)

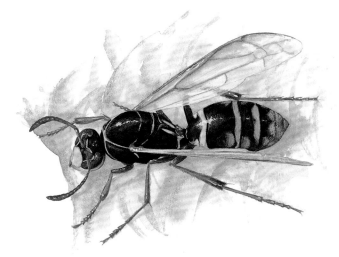

Hornets (*Dolichovespula* spp., *Vespa* spp.) construct the classically terrifying, football-shaped nests that hang in trees and from eavestroughs. The bald-faced hornet (*D. maculata*), also called bull wasp, is a common species closely related to yellow jackets that ranges throughout southern Canada.

Paper wasps (*Polistes* spp.) are the sleekest of all and often resemble yellow-streaked flying missiles. They build smaller nests than their cousins and often favour spots under eaves or in other sheltered locations.

Wasps are actually somewhat beneficial in the garden. They're meat eaters and devour copious amounts of pest insects. They also pollinate flowers, albeit not nearly as well as larger bees with their hairy, pollen-catching legs. If you find a nest that's out of the way and not harming anyone, consider leaving it be. If it's close to the house, however, get rid of it.

Wasps are attracted to sweet, ripe fruit; try to pick it up as it falls to prevent wasps from swarming into your yard for their sugar fix.

CONTROL 1: PREVENTION

Being carnivores, wasps gobble up protein wherever they can find it. In early to midsummer, especially, as they build up strength for the months ahead, they'll be attracted to food scraps, bits of meat, pet food and anything else left out. Be diligent about keeping a clean yard, and they'll find barbecues elsewhere.

By August and into September, wasps will be more attracted to sweets than meats. They love quickly digestible, liquid sugar sources most, so keep your pop and juice covered or just choose water. Wasps have a bad habit of sneaking into pop cans unseen, so use a straw to avoid a very nasty surprise. Also get in the habit of cleaning up fallen fruit from your trees as soon as possible. Wasps will swarm for the free sugar.

Wasps have powerful sniffers, so cut back on the perfumes and fancy aftershaves when you're on a picnic. Most perfumes use floral scents as their foundation, so when you douse yourself in them in order to appeal to that certain someone, you'll actually appeal to a lot of unwelcome others, too.

Try to avoid wearing floral prints and bright colours. Besides not smelling like a flower, you probably don't want to look like one, either. Unfortunately, this suggestion can lead to a tough choice to all you summer sundress lovers out there.

CONTROL 2: ERADICATION

There are a number of chemical-free ways to either control wasp numbers or get rid of the nest entirely, especially if it's exposed. If the nest is hidden, such as in a wall or the ground, your chemical-free options evaporate quickly, and you may need to resort to a killer spray or foam. If you're allergic to stings, use extreme caution no matter what solution you choose.

Hanging traps are very effective at keeping wasp numbers down, and there are several options available that are pretty enough to double as garden decorations. Fill the trap with protein (tuna works best) and not sweet nectar. The latter will attract bees, as well, and we need to keep as many of them in our yards as possible.

If the nest is exposed and you have good reflexes, consider drowning it. As aggressive as they are by day, wasps retreat to their nests at night. Don a long-sleeved shirt and gloves and head out in the dark, large cloth bag in hand. Wrap the bag around the nest, pinching tight at the top, and break off the nest. Submerge the bag in a ready bucket of water and put some rocks over it.

Dish soap and hot water is a surprisingly effective, chemical-free wasp killer. Put 60 mL dish soap in a hose end sprayer or injector filled with hot water. Kit up and head out at night with the sprayer on a hose (the higher pressure the better). Soak the nest until it's covered in soapy water, and if possible, spray right into the entrance hole.

If your fruit tree is attracting too many wasps, use hanging traps to control the pests.

Fungus Gnats

If you have houseplants, then I bet you've had fungus gnats (most commonly *Bradysia* spp.). They are a winged household fixture in late winter, but thankfully, they look like more trouble than they really are.

IDENTIFICATION

Gnat populations tend to peak indoors in fall and late winter. They often arrive as stowaways in fall when we rescue tender outdoor plants from frost.

Old, damp pots of dirt are like new condo complexes to fungus gnats. They live in plants' potting medium (soil) and feed on fungi, algae and decaying leaves in the top 5–7 cm. They ignore the plants completely.

These prolific pests are weak flyers and tend to flit through the air in, zigzagging patterns. They never stray far from home, so when you see one, look for the nearest potted plant as the probable source.

LIFE CYCLE

Reduced sunlight hours in wintery northern climates slow down the metabolism of all the creatures who live in these places, including us and our plants. When our metabolisms slow, we feel lethargic. When our plants' metabolisms slow, they take up less water. Less water uptake means that the top layer of your potting soil stays consistently moist, especially if you dote over your green babies and keep watering at the same tempo as in summer. On top of that, if you haven't repotted your plants in a while, your medium will have degraded; old potting mediums lose drainage and stay wet longer. These conditions are perfect for fungus gnats to live and reproduce.

Like most tiny varmints, fungus gnats reproduce fast and build populations with astonishing speed. It takes about two weeks of gorging on their fungi, algae and rotted leaf buffet before the larvae pupate into adults.

During their approximately week-long lifespan, adult females will lay about 200 eggs in the tiny cracks between soil particles. They prefer peat moss–based mediums, but anything moist will do.

DAMAGE

The good news is that fungus gnats are harmless to you and your plants. They're passionate about the decaying organic matter in soil; the plant being there is happenstance.

That being said, no one wants gnats flitting about during dinner. They're easy to get rid of, and you usually won't require chemicals.

CONTROL

People ask me every day what they need to spray their plant with to kill fungus gnats. I tell them to save their money and just put the watering can away; the secret is drying the plant out.

Water-stressing a plant is never healthy, but if you have gnats, then the plant is probably seeing more water than it needs. Cut back the watering and, when the top layer of soil is completely dry, gently scrape it out, being careful not to disturb the root systems of overgrown tropicals.

Replace the old soil with bagged, sterilized soil that's available almost anywhere. Make sure it's well draining. The new medium will be free of old organic matter and will drain faster than the old.

If you want to be sure, spread a layer of decorative pebbles across the surface.

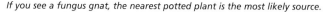

If you see a fungus gnat, the nearest potted plant is the most likely source.

You can also hang yellow "sticky strips" from the plant stems to bring down their numbers.

If the gnats persist after a couple of weeks, reach for a pyrethrum-based insecticide. Spray outdoors if possible and in a well-ventilated room if not. The pyrethrum persists on whatever it touches, so keep pets and children away for a while.

Sprays such as insecticidal soap and neem oil work on contact with adult bugs (not eggs and rarely larvae), so while they are safer than pyrethrum, you'll need to reapply every few days for a couple of weeks to nip the gnats' life cycles.

The best way to control fungus gnats is to water your houseplants less in winter.

Mealy Bugs

Depending on who you ask, this unmistakable critter is either disgusting or fascinating. It looks positively prehistoric. Take a moment to admire them from an evolutionary point of view, then kill them. They're easy to control in small numbers but, if allowed to establish a population, they can be frustratingly persistent.

In warmer climates, mealy bugs (Pseudococcidae) are a financial scourge to tropical fruit growers. In the home, they tend to favour citrus, orchids and big-leafed tropical plants, though they will latch onto almost anything, given the chance. They are most often a nuisance inside in winter, but they can also make their presence known in your summer garden.

IDENTIFICATION

You don't get more recognizable than mealy bugs. The females establish themselves in the crooks of stems or on leaves' undersides and excrete their characteristic frothy white covering that looks like a cottony bundle.

LIFE CYCLE

The females lay their eggs on the undersides of leaves. Look for the cottony white bundles. The eggs hatch in about a week, and the ensuing larvae wander the leaf for 40 days before maturing. Resulting females will lay around 600 eggs in their three-week lifespan.

Male mealy bugs sprout wings at maturation and fly from leaf to leaf calling on females. They are rarely seen but are very busy, and they help make a lot of baby bugs during their three-day lifespan.

DAMAGE

If left alone, the gooey adults will suck enough juice out of your plants' leaves to turn them yellow. Eventually the leaves will drop and the plant, unable to photosynthesize, will die.

With fruit-bearing plants such as citrus, yield will slow and then stop as mealy bug numbers increase.

CONTROL

The trick to mealies, as with other houseplant pests, is to catch them early. Once the white clusters of bugs clumped together emerge, the population is very well established and the plant is in real danger.

Watch for signs of mealy bugs in winter, especially on your citrus (above), orchids (below) and big-leafed tropicals (right).

The mealy bug's frothy protective goo makes it harder to kill but by no means impossible. To start with, isolate the affected plant upon first identification. Although the females can't fly, larvae are known to hitch onto cats and dogs for passage to other plants.

If you catch the population at a few individuals, a strong stream of targeted water, applied every few days, should dislodge them. Make sure to spray the leaves' undersides.

If it's too cold to spray outside and you don't want to drench your living room, spot treatments with rubbing alcohol–soaked cotton balls can be very effective on small populations, as well as on egg bundles. Afterward, wipe the dead bugs, and the alcohol, off with a damp cloth.

Mealy bugs love warm temperatures and will shun the cold. In winter, put your plant next to an open window as long as it can handle a little chill. The bugs will crawl to the leaves farthest from the window, where you can wipe them off easily.

Major infestations will need to be sprayed into submission. Reach for the insecticidal soap first and follow the instructions on the bottle. You'll need to spray every week in order to catch the hatching eggs.

Use a pyrethrum-based spray as a last resort, though if the previous methods don't work, it may just be time to get a new plant. Make sure to keep kids and pets away, as pyrethrum persists on the surfaces it contacts.

Scale Insects

Scale insects (Coccidae) are the clams of the garden, clinging to your plants and protecting themselves under thick armour. They are a problem outdoors in summer, but they also flourish in warm, dry household air during our long winters.

There are over 3000 species of scale in North America. While some species secrete honeydew, which can lead to sooty mould growth, the "clamshells" are the most common. Eradicating them takes persistence, mostly because they typically go unnoticed until there are lots of them, but it's not difficult.

INDENTIFICATION

Their armour is a clever evolution that acts as camouflage as well as protection from predators. Enclosed adults look like natural bumps on the stems (or rough bark on larger plants) until they start overlapping due to numbers, which is typically what sets off the alarm bells.

The nymphs, called crawlers, are too small to look for, but if you develop an eye for the bumps, you'll be able to avoid a major infestation. Watch for legless ovals clustering together; the colour may vary between species, but they're usually about 3–6 mm long.

If you do see strange bumps appearing on your plants, don't assume they're natural. Give them a stiff thumbnail scrape. If they come off easily, they're not part of the plant's anatomy, and you've probably got scale.

LIFE CYCLE

Females, safe under their protective covering, lay eggs, which hatch one to three weeks later. The nymphs search for a suitable place to sink their tiny teeth into your plant and suck the juice out of its cells. This is the only life cycle stage when scale insects are mobile—hence the name "crawlers."

Once the nymphs start sucking, they secrete a protective covering that forms over their body and hardens into a shell. It renders them immobile, but they don't complain; they will live out their days happily sucking the life out of your beloved plant.

DAMAGE

Fortunately, scale will take a while to seriously injure your plant. Left untreated, however, the insects will suck enough fluid to weaken your plant, turn it yellow and eventually kill it.

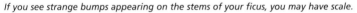

If you see strange bumps appearing on the stems of your ficus, you may have scale.

Inspect your houseplants often to catch scale early.

CONTROL

Like any plant pest, control gets more difficult as the population increases. Wary human eyes are scale insects' worst enemy, so be vigilant and try to catch them early.

When you see them, the first thing to do is isolate the affected plants. Scale insects are invasive and love to spread.

You can easily scrape off the shells with your fingernail or a twig. Don't use knives or other sharp tools that might damage the plant tissue.

There are a variety of tiny insect predators available for purchase. Although often expensive, these critters will wipe out the population and, when there's nothing left to eat, will die themselves.

Chemical cures are tricky because scale insects' shells render the adults invulnerable to most products. The crawlers are vulnerable, but only for a short time, so you'd have to spray repeatedly.

Horticultural oil does tend to work on the adults. It coats the shell and smothers the insect inside by cutting off its air supply. Insecticidal soap will work on the crawlers but tends to be ineffective against the adults.

If you have a major infestation, look to pyrethrum as a last resort. Try to spray this outside your home, and note that although it claims to kill scale, the anecdotal results I've heard have been mixed at best.

Dead bugs aren't courteous enough to plop off and disappear; the vacant shells remain on the plant. So after all the fancy controls, we still come back to scraping them off manually.

Spider Mites

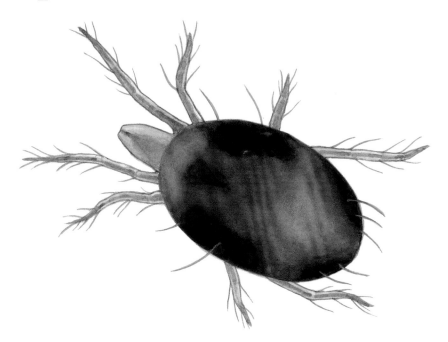

Although spider mites can be a common garden pest in summer, in Canada, they are a winter epidemic inside thanks to our household air getting dry as toast. They thrive in warm, dry places and seem to be a February fixture on houseplants.

A member of the mite family (Tetranychidae), they get their nickname from the protective webbing they spin. While they prefer some plants over others, they will do serious damage to anything they infest. The good news is that, with persistence, they are relatively easy to control.

In the yard, spider mites have many natural enemies, including ladybugs, which usually keep their numbers in check (except they tend to pop up during long, hot droughts), but indoors, we need to control them manually.

IDENTIFICATION

These nasty critters could easily be called vampire mites. They feed by puncturing surface plant cells and sucking the juices right out of them.

Like vampires, they shun sunlight whenever possible and attack the undersides of the leaf first. By the time mite damage is obvious, it's extensive, so try to catch it early.

Spider mite damage

Early signs include yellow discoulorations on the leaf surface. Turn the leaf over and you'll see tiny (less than 1 mm across) mites bustling about, and with a magnifying glass, you'll be able to see their perfectly round eggs scattered randomly.

Signs of an extensive infestation are clumps of webbing around the leaves, rusty red spots turning to brown and scores of creepy specks moving across the webbing doing their dastardly work.

LIFE CYCLE

When it comes to population booms, these varmints know all the tricks. Their eggs can hatch in three days. Offspring are sexually mature in five days, and females will lay 20 eggs a day for their four-week lifespan.

DAMAGE

Spider mites attack a wide variety of plants, especially vegetables and citrus. As they suck the leaf cells dry, the leaves will droop, then drop, and eventually the plant will die.

CONTROL

As bleak as the reproduction math is, there are solutions. I recommend against reaching for a chemical cure unless the situation is dire. Instead, imagine the tropics, and the relaxing sound of a warm, heavy afternoon rain splashing over broad, tropical leaves. That rain is doing much more than providing white noise for napping; it's a natural preventive for

Spider mite webbing

spider mites, which thrive on the dry tissue of partly desiccated leaves.

Just as dry leaves (i.e., when you've been watering only the soil) attract spider mites, moist leaves repel them. When you see spider mites, all you need for your first offensive is a good spray bottle.

Spray the underside of every leaf vigorously with as strong a jet as you can get without damaging the leaf. Pay particular attention to the crooks where the leaf stem grows out of the main stem (the nodes). Lay some plastic around the plant's base; you're going to be spraying water and, hopefully, knocking scores of mites off the leaves as you go.

Take your time, and remember they covet shadows, so get under every leaf.

Hurrying things along will lead to missed mites and a soon-to-be rebounded population.

With this in mind, regularly misting the leaves will help keep a population from getting established, although it won't do much against a raging outbreak. Regularly soaking your plants' leaves down is ideal but also unrealistic in homes.

Spray every few days and, if that's not working, it's time to try something more potent. Grab some insecticidal soap and spray it on (ease up on the jet spray this time). The soap is safe for you but clogs their pores; you're basically suffocating them.

Repeat the soapy spray every three to five days to make sure you catch the

mites right after they've hatched and before they breed. The key to beating spider mites is to match your spraying to their life cycle, and to be stubborn about it for a couple of weeks.

If the soapy water still doesn't work, look to neem oil or a pyrethrum-based spray and follow the same cycle. If using pyrethrum, make sure to get all the bugs; their short life cycle allows them to adapt easily to pesticides, so if you use products such as pyrethrum but don't kill them all, the survivors will develop resistance over time and magnify your troubles.

You can turn to predators, which are readily available online, for an effective and organic solution. They can be pricy, however, and involve introducing another insect into your home.

If your houseplants get spider mites, be persistent in spraying them every few days to kill them all.

Fruit Fly

We've all been there. You're enjoying a ripe bunch of bananas and fresh summer peaches one day and they're buzzing with little flies the next. Fruit flies (*Drosophila melanogaster*) are magicians, appearing seemingly out of nowhere to transform your fruit bowl from delightful to disgusting in almost no time at all.

Although we rarely notice them in the garden or on plants, I'm including fruit flies for their sheer ubiquity, ick-factor and surprising ease of control. We'll never be permanently rid of them, but at least we can learn how to get rid of an outbreak in the home.

FRUIT FLY 101

Fruit flies appear so suddenly, and so alarmingly out of nowhere, that it's easy to believe they reproduce via spontaneous generation. Also called vinegar flies or wine flies, fruit flies' rapid speed life cycles enable their populations to explode as long as they have food. They can go from egg to adult to egg again in a week and can lay 500 eggs at a time. When you see one, the clock is ticking.

Having fruit flies doesn't mean your house is filthy. The fact is that they have an exceptional sense of smell and can sniff out ripe fruit from a long way away. When fruit gets overripe and the smell intensifies, it's a fruit fly dinner bell. They are small enough to fit

Fruit flies are attracted to overripe fruit; if you leave ripe fruit out too long, fruit flies are sure to follow.

through screens or door cracks. You might even have brought some eggs home with you on that banana bunch. Don't stress about where they came from, and don't waste energy trying to ensure they don't come back, because they will. Focus on simple ways to get rid of them.

CONTROL

Fruit flies will always be attached to overripe fruit, so as soon as you see the flies, find the fruit in question and dispose of it. Wash the bowl well to clear out any remaining eggs. Check all remaining fruit and dispose of anything with cuts or open wounds. Empty your indoor compost bin, as well. Put it all in a sealed bag and throw it in the outside bin.

After that, wipe down your kitchen garbage can (the interior, too) with vinegar.

They need a surprisingly small smear of fruity goo to be able to reproduce. Wash your dishes and check for old wine glasses around the house. They love the leftover wine in the bottom.

They're sneaky and like laying eggs in unexpected places. Develop a habit of keeping the stopper in your kitchen sink, as the drain is one of their favourite and most frustrating spots. They can also breed in dirty sponges, so if that thing in your sink doesn't pass the smell test, out it goes.

Fruit fly traps are easy to make and surprisingly effective. Grab a mason jar and drop in some ripe banana and a splash of apple cider vinegar. Snug some plastic wrap around the top and poke 1 or 2 very small holes. The smell will lure them in, and they'll never come out.

A pot of basil near your fruit bowl will help keep fruit flies away.

You can also leave out an open wine bottle with an inch left at the bottom (though I'd rather drink it). Create a paper funnel at the top with as small an entry as possible.

Once that infestation is cleared, there are steps you can take to prevent another one. Fruit flies hate basil, so keep a pot of it next to your fruit bowl to deter them (and you get fresh basil as a bonus).

Disgusting Little Heroes

Even as I try to get rid of them, I tip my hat to the lowly fruit fly. Over 100 years ago, Thomas Hunt Morgan began incorporating fruit flies into his explorations into genetics. Their quick life cycle and adaptability, combined with their surprising genetic similarity to humans (humbling, I know), make them the ideal test subjects. Fruit flies helped Morgan win the 1933 Nobel Prize for his groundbreaking work in genetic inheritance.

Since Morgan, fruit flies have enabled researchers to do everything from charting our genetic map to exploring the relationship of genetics and cancer. Today, thousands of labs across North America are using fruit flies to understand the genome, cure disease and improve quality of life for millions of people. The next time you shoo them off your bananas, give them a nod.

Fairy Ring

While most mushrooms that pop through your grass during wet summers are harmless, some are aggressive enough to damage your lawn. Fairy rings have always been a sore spot for Canadian gardeners because they are highly visible and devilishly difficult to get rid of.

Fairy rings have a rich folkloric background and have been called elf rings, witches' rings and sorcerers' rings in Europe. Their tendency to occur in woodland areas have linked them with supernatural stories of fairies and other elusive creatures.

Western European oral traditions are brimming with stories of fairy rings appearing as gateways between our world and the kingdoms of elves and fairies. The sprites would appear and dance within the ring, a space inhabiting both worlds, until they went back to their own world, taking the ring with them.

Keep in mind that lawns weren't a big thing for medieval Europeans, and fairy rings usually occurred in the woods, which were rife with their own mysteries. Folkloric stories promised that if you waited at these rings for the elves to return from their earthly mischief, you could catch them on their trip home.

IDENTIFICATION

When you see mushrooms, you're essentially seeing only the "flower" of the "plant." The rest of the fungus lies beneath the surface, and it only sends up the mushroom cap to spread spores around the neighbourhood.

The rings themselves are impossible to miss. Look for giant horseshoe-shaped patches of dead grass, usually 1 m wide. A heavily infested lawn will have several of them in close proximity, but they will very rarely overlap.

LIFE CYCLE

Fairy rings are caused by several species of mushroom that naturally occur in arcs. Like all mushrooms, the fungal fibres (mycelia) are mostly underground. It spreads via microscopic filaments called hyphae, but also via not-so-microscopic lawnmowers as they cut and chuck bits of the emerging mushrooms all over the lawn.

DAMAGE

Unlike most mushrooms, fairy ring mycelia get so dense that the lawn can't get any moisture, which creates a necrotic (dead) zone in your lawn and looks awful.

While fairy rings won't technically kill the whole lawn, they will eventually overtake your lawn until it looks terrible.

Even when you don't see the actual mushrooms, the effects of fairy ring are obvious.

To control fairy ring, pick the mushrooms as soon as they appear.

CONTROL

Evicting fairy ring from your yard is no easy task, which is why it's proven to be such a gardening headache.

The first thing to do is pick the mushrooms as soon as they appear; you don't want them to go to spore. Use a mowing bag if you mow over the mushrooms.

The grass above fairy ring dies of thirst because the fungus repels water away. To counter this effect, jab a deep-root feeder across the ring every few inches. Make sure it gets at least 30 cm deep. Soak the spot well twice a week, and you should drown out the fungus. If you don't have a deep-root feeder, a pitch-fork will suffice, although it won't get as much water down deep.

Add 5 mL dish soap to 4 L of water and pour it on the affected area. The fungus doesn't like nitrogen, so hitting it with a high-nitrogen fertilizer will also help.

The only other solution that I know of involves digging out the entire affected area, replacing it with fresh soil and starting the lawn from scratch. How deep you have to dig depends on how deep the fungus is. You'll need to dig even a few inches below the last visible strands. I've heard of digging to depths of 15 cm all the way to 1 m to dig out all the white mycelia (you'll know them when you see them). Make sure to dig 30–45 cm out from all sides of the ring, as well.

Other Mushrooms

During wet summers, people come into the greenhouse every day asking what they can use to get rid of the mushrooms sprouting in their lawn. The good news is that mushrooms won't harm your grass and will disappear as moisture levels fall. The bad news is that, other than snipping off the toadstools, there's nothing you can do.

Like bacteria in your body, there are fungi throughout every healthy yard even though you see only a fraction of them. Fungi live off rotting organic matter and, in the process, convert much of it to nutrients for plants to use. So while they're not the sexiest group of organisms around, they're actually one of the most useful.

IDENTIFICATION

Let's face it, there are always fungi growing beneath your lawn, especially if it's older. The most common mushrooms you'll see are the simple grey or white capped mushrooms. No matter what type you find, resist the temptation to eat them unless you know exactly what you're doing. Toxic species exist widely in Canada.

LIFE CYCLE

Think of the visible portion of a mushroom as a flower. Below it, there's a vast network of microscopic filaments, called hyphae. When hyphae bind together to form a mass of white or dark, thread-like growth, it's called a mycelium, which is what you would find beneath that humble toadstool.

Just like a plant waits for the right conditions to flower, the mycelium waits for sufficient moisture to "bloom" (send up mushroom caps). During dry years, the fungus simply lies dormant under the surface.

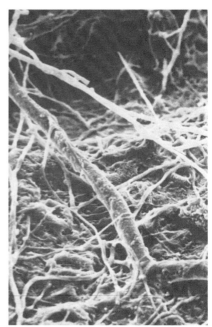

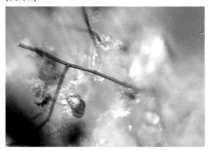

Mass of fungal strands (above); fungal hyphae (below)

DAMAGE

Although mushrooms aren't the most attractive addition to the lawn, they are doing much more good than the aesthetic harm. Fungi act as an underground composter of organic matter, which would otherwise accumulate endlessly.

CONTROL

The white mushrooms in your lawn are feeding off old thatch, buried bits of wood and other assorted decaying matter that the mycelium is wrapping around. If you have an older lawn, annual aerating to clear out the thatch would be good start to reducing their numbers.

Surface mushrooms' only goal in life is to spread spores. To keep them in check,

---

Fun Fact

The visible part of a fungus might look small, but when filaments keep threading together, they can form massive organisms. A mushroom in eastern Oregon is often described as the largest living thing on earth and covers 2200 acres.

Use an aerator once a year to help reduce mushroom numbers in your lawn.

try to pick them before they go to spore, as each of the countless spores they release into the air will become hyphae of their own and will seek to build a new underground infrastructure.

When you mow your lawn, you may want to use a bag if mushrooms are widespread. Collecting the shredded fungus bits in a bag will keep them, and the developing spores, off the lawn.

If you think you're willing to do anything to keep mushrooms off your prize lawn, think again. There's no chemical treatment, and the only way to make sure they're gone is to dig them out. You'll need to dig up the entire yard down deep enough that you can't find any more white strands of mycelium—often up to 1 m deep. Even then, there will probably still be hyphae that you don't see, which will promptly resume growing, weaving and getting thicker until one wet summer day, when you'll look out the window and get an unpleasant surprise.

My advice is to learn to live with them, and if you're so inclined, take the opportunity to learn about fungi and the amazing role they play in our ecosystem.

Can I Eat Them?

When people ask me if they can eat the mushrooms in their lawn, I tell them if they have to ask, don't eat them. While we have many edible (and downright delicious) species in Canada, we also have toxic species that can and have killed people.

If you want to learn about eating mushrooms, talk to someone who can take you out and show you what is edible and what isn't. No matter how innocent they look growing in the yard, don't go by pictures on the internet or from a book alone to identify them.

Powdery Mildew

If spring is cool and wet, expect to find some type of fungus popping up across the yard. Much more than mushrooms, fungus can be powdery mildew on lawns, black knot (see page 115) on mayday trees, or rust (see page 110) on hollyhocks and other plants.

On lawns and in gardens in areas with poor air movement, powdery mildew (Erysiphaceae) is a common result.

IDENTIFICATION

Powdery mildew looks like a white powder sprinkled on the lawn or on plant leaves. It starts out as tiny spots of white on the leaves. They get bigger and more spots appear until they merge and become an ugly white blob covering most of the leaf.

Watch the parts of your garden or lawn that are sheltered or perpetually shady, such as the area between houses. If the foliage or grass blades stay wet overnight, when the air cools, fungus is likely to set in.

LIFE CYCLE

There are a number of different species of powdery mildew, but its life cycle is always the same. As a fungus, it reproduces via spores produced in the white patches; the wind disperses the spores, as do gardeners who aren't careful when they prune or mow diseased areas.

Powdery mildew survives winter on infected plant debris piled in the garden. It actually produces structures on the leaves, appearing as little black dots, which shelter it from the cold and help it survive.

The good news is that every species affects a specific host. If one kind of plant in your yard has it, that mildew will spread only to other individuals of the same plant.

DAMAGE

A host of different plants are susceptible to different species of powdery mildew. Whether wheat, grapes, onions, apples, melons, lilacs, strawberries or grass, the attacking species is different, but the damage is the same. The spots start on the leaves and can eventually creep up the plant to infect all of it.

The mildew is unsightly, but it probably won't kill your plants or lawn. Because the mildew reduces plants' ability to photosynthesize, serious cases can lead to decreased yield and a weaker plant overall. A weaker immune system can make the plant vulnerable to other diseases, so in that way the mildew can kill indirectly.

CONTROL

Most of the perils of wet weather in flower beds can be controlled by removing old plant matter, weeds and any other debris. When rotting leaves (possibly left over from winter), weeds such as chickweed, and other junk clutter the spaces between your perennials, the air flow is inhibited, and the foliage stays wet. This leads quickly to mildew.

Improve the air circulation if possible by pruning, and definitely don't water. Remove any diseased plant material. If the problem persists, you may need to apply a sulphur- or copper-based fungicide. Never apply more than the label recommends; doing so won't get rid of it faster and may cause more problems.

Milk is a surprising home remedy. Diluted in water (about 1:10), it's anecdotally as effective as many store-bought, often environmentally harmful fungicides. Spray the mixture on leaves either at the first sign of infection or weekly as a preventative measure.

Cleaning up your yard in fall will reduce the likelihood of powdery mildew surviving winter.

Snow Mould

Wouldn't it be nice if we were greeted by green, fresh grass when the snow finally melts? Unfortunately, the thaw often leaves behind yet another yucky mess. If you find your yard covered in snow mould after the big thaw, don't despair. It won't last long, and there are ways to prevent a similar outbreak next spring.

IDENTIFICATION

Also called Typhula blight, grey snow mould (*Typhula* spp.) produces 7–30 cm wide, circular patches of silvery grey fungus that looks similar to spider webs on your lawn. As allergy sufferers know, it can wreak havoc on the sinuses, but usually it's otherwise harmless and will clear up quickly as the lawn dries.

LIFE CYCLE

Like all fungi, snow mould happens when dormant spores are given the right conditions to thrive. The time to think about snow mould is in fall; by the time the snow falls, the severity of your spring outbreak has been largely decided.

If the winter snow cover arrives before a cold snap, more spores will survive (along with more mice) in the warmth trapped under the snow. During the thaw, melting snow provides the moisture necessary to energize the dormant spores.

The mould will vanish as soon as the lawn dries. If there's a lot of snow, and if there are significant spring rains, this may take a week or more. Once fall rolls around, focus on prevention.

DAMAGE

Grey snow mould infects grasses' leafy tissue and causes only superficial, indirect damage. If it's thick, however, it will choke out the lawn beneath it and leave bare or brown patches in its wake.

CONTROL

To prevent the mould, avoid high-nitrogen fertilizers late in fall; reach for the fall blend instead. Keep cutting the grass until it stops growing and, especially if you have a mature lawn, rake and aerate it to keep thatch at a minimum.

Fungus hates wind, so if you have big trees rimming your yard, thin out any deadwood. On top of keeping the snow mould down, pruning will give your trees a healthy boost.

Once the mould appears, the only way to get rid of it is to wait. If you're an allergy sufferer, you'll want to close your windows. You may be tempted to apply a fungicide to get rid of it, but I don't recommend spending the money or the energy. The mould doesn't last long, and if there's going to be any damage to the lawn, it's probably already happened by the time you notice it.

Once the lawn dries out, rake up the dried fungus and dead grass and throw everything in the garbage (not in the compost pile). Make sure to disinfect your tools once you're finished.

Wait until the lawn is actively growing and simply sprinkle a bit of sterile soil and some lawn seed over the affected areas. Water, wait, and it will be like the mould never happened.

Your lawn will usually make it through an outbreak of snow mould unscathed.

Rust

Seeing as how it makes healthy leaves look like the bumper of an old Chevy, rust (Pucciniales) is aptly named. There are over 7000 species of fungal rust worldwide, and outbreaks are able to wreak havoc on everything from wheat to coffee crops. In 2014, virulent outbreaks of striped wheat rust were reported across the Prairies, and headline-making coffee rust is threatening coffee drinkers' fixes worldwide.

IDENTIFICATION

Unlike most pathogens, which favour immature growth, rust will most likely appear on mature, healthy leaves first. Look for it during years when the rain keeps falling and, especially, when the evenings cool off enough for wet leaves to stay wet and cold all night.

Check mature leaves for flecks of rust; it looks just like actual rust on a car. It will probably appear on the undersides first. The flecks will turn bumpy and spread until it looks like the leaf has been turned upside down and sprinkled with fuzzy, orange paint.

Hollyhock rust (*Puccinia malvacearum*) is common in residential yards. You might find similar rust on asters, carnations, geraniums, irises, lilies and pansies.

LIFE CYCLE

The fungus will overwinter on fallen infected leaves if they aren't tidied up before the snow flies. In spring, if you haven't picked up leaf litter around infected plants, the spores will wake up and start spreading.

DAMAGE

Although rust rarely kills the infected plant directly, it can weaken it enough to reduce yield, deform leaves and make the plant vulnerable to gall and other diseases. In hollyhocks, once the rusty red pustules turn a chocolatey brown, leaf drop is imminent, followed by stunted growth.

CONTROL

If the rain is falling, put away the hose. Rust thrives with excessive moisture. Clearing out dead wood and pruning inner leaves to increase air circulation will help to dry leaves out, and dry leaves have less chance of being infected. When you do water, keep the hose low so the leaves don't get splashed.

Remove and throw away infected leaves as soon as you see them (put them in the garbage, not the compost). Sprinkling sulphur dust or applying neem oil on affected leaves will slow the spread, but it will be easier and most effective to remove them. Make sure to sterilize your tools before you move on to other plants.

If you choose sulphur dust or neem oil, you'll need to re-apply every 10 to 14 days throughout the growing season, which can get pretty onerous. Establishing good watering habits, thinning out centre growth and practicing better fall sanitation will be less time consuming and more effective.

Both hollyhocks (above) and irises (below) are susceptible to rust during cool, wet summers.

Juniper-Hawthorn Rust

While wet weather makes for lush gardens, it can also create the perfect conditions for some bizarre and unsightly garden problems. If you've seen alien-looking, orange-tentacled blobs sitting in your junipers or cedars, you know what I'm talking about.

Concerned gardeners are bringing scores of samples and stories of juniper-hawthorn rust (*Gymnosporangium* spp.) into the greenhouse daily. It's a specific form of fungal rust that, while not fatal to infected plants, does make them look like they've been "slimed."

Cedar-apple rust on its primary (above) and secondary (below) hosts

IDENTIFICATION

If you or your neighbours have had the disease, look for a brown-orange gall on your evergreen stems in late summer.

If you miss the galls and your junipers are infected (or your cedars with cedar-apple rust), you will notice gelatinous orange blobs hanging off the branches the following June. They look like orange octopuses, with slimy tentacles (called teliohorns) dangling from the main gooey mass.

To prevent juniper-hawthorn rust, avoid planting junipers (above) anywhere near hawthorns (below).

The secondary hosts don't acquire the gelatinous teliohorns. They get rust spots on their leaves that are orange with dark red to black centres.

LIFE CYCLE

Juniper-hawthorn rust is closely related to cedar-apple rust; the only thing that really differentiates them is the plants they infect. They need two specific hosts to complete their two-year life cycle.

The disease starts as a gall on juniper or cedar stems in late summer. The slimy, orange teliohorns won't emerge until the next spring.

Each teliohorn has thousands of spores inside of it just waiting to catch a ride on the breeze and travel through the garden. When they mature in mid-summer, the dispersing spores can only latch onto a specific secondary host.

In the case of juniper-hawthorn rust, the spores travel from their primary host (junipers) to their secondary hosts (hawthorn, mountain ash and apples). With cedar-apple rust, the spores travel from cedars to apples, mountain ash, pears and saskatoons.

In late summer, a different kind of spore blows from the secondary host back to the primary, where it lurks over winter until it can re-emerge in spring.

DAMAGE

The good news is that this disease, though disgusting, won't cause significant damage to your junipers. The bad news is that if it hits your apple trees hard, it can lead to significant defoliation and a reduced fruit yield.

CONTROL

Whether you can catch the growing gall or find the unmistakable orange blob the next spring, it's important to remove the growth before it releases the spores and damages the secondary hosts. Cut all infected branches off 20 cm from the growth. Burn the fungus and disinfect your tools with a light bleach solution.

If it's already on your apples, hawthorns and mountain ash trees, remove as many infected leaves as feasible. The fewer leaves there are, the fewer spores will find their way back to the primary host to continue the life cycle. As a precaution, if you find the orange blobs, you could sprinkle some sulphur dust on the leaves.

The best prevention is to keep the primary and secondary hosts separate. If you're in the midst of planning your yard, try to leave some distance between these varieties.

I consider this fungus well established. While you'll see more of it on rainy years and less on dry years, it's here to stay, so the best offence you have is to learn how to identify it early and keep it off your deciduous trees.

Mountain ash (above) and apple trees (below) are possible secondary hosts for both juniper-hawthorn rust and cedar-apple rust.

Black Knot

All across Canada, black knot can explode to outbreak proportions with astonishing speed. This easy-to-control fungus destroys garden aesthetics, disfigures branches and eventually kills trees.

Black knot (*Apiosporina morbosa*) afflicts trees in the *Prunus* genus. This includes plum, apricot and cherry trees, but it's most aggressive on maydays and Schubert chokecherries.

Black knot is nothing new. It was described in Pennsylvania almost 200 years ago and wreaks occasional havoc with commercial fruit crops in North America. It's a natural disease, but our love of the *Prunus* genus has given it the opportunity to become unnaturally invasive.

IDENTIFICATION

Whether driving or walking, you've probably seen black knot marring tree limbs. It's startlingly evident in winter, when the bare limbs look like they have charred masses of burnt rope wrapped tight around the branches and trunk.

A big reason that black knot spreads so quickly is that it doesn't assume its classically nasty, charcoal look until after it's gone to spore and infected the trees around it. Before that, it's easy to overlook.

LIFE CYCLE

Black knot spreads fastest during warm, wet springs. If you're vigilant, you'll see a brown swelling on this year's or last year's growth. It will look as if the branch was stung by a bee and swelled up.

Fast forward to the following spring. Now the fungus grows into a bulbous, olive green knot. At this point, it gets nasty.

Around the time when the tree leafs out, black knot goes to spore, with the number of spores released typically peaking as the tree blooms. The more spring rain we get, the more spores we get, which the wind promptly carries to neighbouring trees.

Spores may emerge again in fall if it's wet enough, but our autumns tend to be dry. Typically, the fungus needs six hours of near-steady rain in order to go to spore.

DAMAGE

Black knot causes damage by turning trees' own branches against them. The fungus causes rapid growth of plant cells until they distort and stunt the tree. Left unchecked, the knots will spread lengthwise every year, slowly devouring their host.

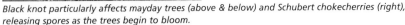

Black knot particularly affects mayday trees (above & below) and Schubert chokecherries (right), releasing spores as the trees begin to bloom.

CONTROL

The only reliable treatment is pruning. All infected growth needs to be cleanly cut off 15–25 cm down from the knot. If the infection is widespread, this often means taking out the tree. It's a tough task, but necessary for the sake of your, and your neighbours', other trees.

As you prune, keep a bleach solution nearby to sterilize your cutters after every cut; otherwise you may be leaving trace amounts of spores on every branch the blades touch.

Treat infected branches like hazardous waste, as they can spit out spores up to four months after pruning. Burn or throw them away—do not compost them. If you put them in the trash, wrap the knotty bits in a plastic bag.

Never prune out black knot in spring or in wet, warm weather. The ideal time is between early November and the end of March, when the tree is dormant and the lack of leaves makes for easy navigation.

Even if you don't have black knot, prune your trees regularly to clear out old wood and to keep the air circulating in and around the tree. Regular pruning keeps branches drier and less hospitable for fungus.

Fungi thrive in moisture. Don't water your yard when it doesn't need watering. When you do water, try using a drip hose or another tool that keeps moisture close to the ground instead of spraying it into the air (and the trees); it will help you conserve water as well as avoid black knot.

If you have a lot of deer, porcupines or other critters nibbling your tree bark, wrap it tight over winter. Black knot spores love finding homes in fresh, moist scar tissue.

If your neighbours have black knot, talk to them about its seriousness in the hopes that they will do some pruning. Bad tree diseases don't always make for bad neighbours; sometimes they are the common cause needed to bring neighbours together.

Fire Blight

Some of the worst diseases Canadian gardeners face come from fungi. While some spores take years to do permanent damage to a tree (black knot, for example), others can lay waste to an entire apple orchard in a season and leave nothing but charred decay in their wake. I'm talking about the spores that lead to fire blight (*Erwinia amylovora*).

Like most fungal diseases, fire blight is most virulent during wet, humid growing seasons. Its favourite prey is apple and pear trees, but it also attacks crabapple, hawthorn, cotoneaster, raspberry and mountain ash. Once it's established, the only cure is to destroy the host, so the focus needs to be on prevention.

IDENTIFICATION
see Life Cycle

LIFE CYCLE
Fire blight is a wasting fungus that shows different symptoms as it progresses through the tree. For simplicity's sake we'll start with the lovely blossoms of spring. Typically, pollinating insects unwittingly spread the fungus after indulging in the sticky, sweet ooze coming from an infected tree's cankers.

The insect transfers the fungus to an open flower, and shortly thereafter, the bloom begins to look grey-green and then black. This is called blossom blight, and when the petals drop, the fungus creeps up the branch to became shoot blight. The tender branch turns brown and wilts in a characteristic candy cane or shepherd's hook shape. This is usually when we realize, to our horror, that our tree has contracted fire blight.

If it's a warm, humid summer, look for your first glimpse of sticky droplets of disgusting ooze at this point. If it's rainy, expect the fungus to spread rapidly toward the older growth at the centre of the tree. As it spreads, infected wood will look like it has been scorched by flame—hence the name.

As the infection sets in, cankers form in the older wood. The cankers, which are the third and most devastating phase, secrete bacterial ooze and form open wounds that, in turn, fresh fungal spores can colonize to hasten the tree's doom.

Flowering pear (above) and apple (below) trees are particularly vulnerable to infection.

The fungus spores overwinter inside the cankers and wake up festering with the first rays of spring. They start to ooze just in time for the bees to carry their sticky poison to opening blossoms, thus beginning the nightmare anew.

DAMAGE

While infected fruit trees will usually still yield, the fruit will be small, shrivelled, may have nasty lesions and will probably be dripping with sticky goo. Left unchecked, the fungus will almost certainly kill the tree.

The disease spreads rapidly when trees are under stress and are sporting open wounds; it's been known to infect entire orchards after one punishing hail storm.

CONTROL

With no effective cure, the strategy needs to focus on prevention. In spring, check your apple and pear tree flowers for any greyness or discolouration and, if found, remove them immediately. After that, if the young shoots start to wilt into a shepherd's hook, prune them off immediately, including a foot of healthy wood as buffer, and disinfect your tools afterward.

Wet springs are a boon for all fungi. If the May rains don't relent, be vigilant. Don't give the garden any excess water, and try to prune out dead wood in order to increase air circulation (which, in turn, dries out wet, porous wood faster).

Not all fruit trees are created equal. Before you buy your apple or pear tree, do a quick internet search of what cultivars are most susceptible to fire blight, and try to avoid them. Increasingly, hybridizers are focusing on bringing varieties to market that have some resistance to fire blight.

Washington hawthorn is a species resistant to fire blight.

Bronze Leaf Disease

Do you have any Swedish aspen in your yard? If you live in a newer suburban home and have a smaller yard, you probably do. They are the tall, columnar trees you see lined up against backyard fences. In just a decade, they've skyrocketed in popularity as (especially western) Canada's favourite space-saving tree. But they aren't without potential problems.

Bronze leaf disease (*Apioplagiostoma populi*) is a fungus that infects trees in the poplar family, namely Swedish aspen, trembling aspen and tower poplars. Pervasive and virulent, the fear is that it will spread from ornamental Swedish aspen to our native stands of trembling aspen, which constitute much of western Canada's native forest and urban forest.

IDENTIFICATION

The disease announces itself in mid-summer when sections of leaves (although sometimes they're scattered) turn anywhere from bronze to chocolate brown.

By late summer, the discoloured leaves will start to dry and curl inward. Note that the leaves' veins often remain green until then, which is how arborists distinguish this disease from other leaf-browning blights.

In fall, infected leaves will cling creepily to their branch as others flutter down. If you didn't see the infected leaves in summer, they will be painfully obvious now.

Watch for signs of bronze leaf disease on your Swedish aspens.

LIFE CYCLE

Although the entire tree is infected by the time you see the discoloured leaves, these are the leaves capable of spreading it.

In spring, infected branches' leaves may unfurl either green or yellowish green, but they will be stunted in size. Once daytime temperatures rise above 18°C, the spores will take flight and infect everything around them.

DAMAGE

Bronze leaf disease a nasty bit of business, with no chemical treatments and the ability to kill your tree in three to five years. The tree's branches become deformed, and the brown leaves aren't able to photosynthesize, so ultimately, the tree slowly starves.

CONTROL

There's no chemical treatment for bronze leaf. It lives inside a tree and, once established, can't be eradicated. The only treatment is containment via removing affected leaves and branches or, as a last resort, the tree itself.

Any discoloured leaves must be removed and garbaged (not composted) right away. Removing infected leaves as soon as you see them will extend the life of your tree and, hopefully, save others nearby.

It's good practice to gather an infected tree's other leaves in fall and garbage them to prevent the fungus from potentially overwintering on the ground.

Cut off affected branches 20–30 cm back from the last visible sign of infection (either an infected leaf or an area browned under the branch's bark). You may be throwing aesthetics out the window by doing this, but you'll be saving your tree from an untimely end. Make sure to bleach your pruners after cutting.

Healthy trees are better able to repel diseases than sickly trees. Make sure to plant your aspen in good soil and keep it watered and fertilized.

If you don't want to take the chance with aspen, there are may other columnar options out there for smaller yards, including columnar crabapples and mountain ash. The latter will draw flocks of cedar waxwings to the fermented berries in spring.

Tower poplars are also susceptible to bronze leaf disease.

Dutch Elm Disease

Originating in central Asia, the mighty elm is one of the world's most versatile and stately trees. The Persians were the first to prune elms and use them ornamentally, and in the 19th century, as Europeans began investing in parks, urban forests and boulevard trees, the elm became ubiquitous in cities everywhere.

Sometime during World War I, European elms started dying. People didn't know why at first, with many blaming traces of nerve gas from the Western Front. In 1921, a Dutch scientist named Bea Schwartz isolated a fungus from the wood of dying elms. It was *Ophiostoma ulma*, and it acquired the common name of Dutch elm disease (DED).

Around 1931, an Ohio-based furniture company imported infected elm wood from France, and DED had its North American beachhead. It arrived in eastern Canada during World War II and has since spread to almost every province.

In their native Asia, elms had developed a natural immunity to DED. European and American elms, however, had never been exposed. The impact of the disease on cities lined with hundreds of thousands of aging elms was devastating. Britain has lost more than 25 million trees, and France has lost 90 percent of its elm forest. In North America, the disease has killed about 75 percent of an estimated 77 million pre-DED elms.

Proudly, Alberta has the largest population of healthy elms in the world, with over 200,000 trees (worth over $600 million) in our urban areas. Even with frequent elm beetle sightings across the province, the only confirmed case of DED was a tree in the town of Wainwright in 1998, which was immediately destroyed.

IDENTIFICATION

The first visible signs of DED appear in late June to mid-July, when you'll notice leaves on one or more branches wilting. They will turn brown throughout summer but will remain on the tree. Later in summer, infected trees' leaves will yellow and fall off earlier than they naturally should.

DED relies on elm bark beetles to spread it. Bark beetles are a group that includes the mountain pine beetle currently ravaging western North America. They're called bark beetles because they reproduce inside trees' inner bark.

The beetles are hard to find, which is why inspectors often use yellow sticky strips to trap them. Given their small size (less than 5 mm long), their entry holes into the tree are tiny. Watch for trace amount of sawdust caught in the bark or around the base of the tree. If you strip the outer bark away from an infected tree, you'll find meandering tunnels that the beetles have chewed through the tree.

Early signs of DED damage

LIFE CYCLE

An elm bark beetle burrows under the tree's bark and infects the tissue with the fungus as it chews the soft interior wood for food.

Once the beetle introduces the fungus, it spreads rapidly through the tree, both vertically into the crown and down into the roots. If the disease catches the tree's sap stream (basically its circulatory system), tree death can happen in one to two years.

DAMAGE

A lot of people don't think of plant diseases as having the ability to be devastating on a global scale. Dutch elm disease proves them wrong. In the last century, this disease has devastated rural and urban forests across the Northern Hemisphere and forced us to dramatically re-think our approach to urban forestry, pest management and conservation.

CONTROL

Like most diseases, prevention is the best defence, followed by early detection.

If you believe you have an infected tree, contact your municipality immediately so that they act quickly to stop the disease in its tracks.

Because elm bark beetles are carriers, efforts to prevent and contain DED are largely focused around tracking and eradicating the beetles. The best weapon that municipalities have against the beetle is a chemical called Dursban, which is controversial enough to be banned for household use. Consistent applications can

The susceptibility of American elms (above & below) to DED has devastated urban and natural forests across Canada and the United States.

Chinese elm (above) has natural resistance to DED; new DED-resistant varieties of American elm (below) are being continually developed.

extend a tree's life but, unfortunately, often even that won't save it in the long term.

It's important never to prune an elm tree during the growing season (from March to October). During this time, the running sap in open wounds makes the tree more vulnerable to opportunistic elm bark beetles. Many Canadian cities will fine people who prune elms during the growing season.

Scientists around the world are busy hybridizing and cloning DED-resistant strains of elm, but while the dream of creating a new generation of healthy urban elms lives on, the fungus has a history of adapting itself into ever more virulent strains. It's a worldwide arms race that shows no sign of slowing down.

Late Blight

From 1845 to 1852 in Ireland, approximately 1 million people starved to death as the potatoes they relied on rotted in the fields. Like many tragedies throughout history, the horror of the Great Famine arose from what is now a common and very treatable problem.

Late blight (*Phytophthora infestans*), also known as potato blight, is a nasty fungal disease that rears its soggy head during wet, cool summers. It can easily ruin your entire potato crop, and it can affect tomatoes and eggplants, as well.

The good news is that late blight is easy to spot and contain, but the bad news is that the infected plant will die within days.

IDENTIFICATION

Late blight starts on a plant's leaves. Watch for brown, soggy spots on leaves, stems and fruit, in the case of tomatoes, that spread quickly in moist weather. By the time you see it, it's too late to save the plant, but it may not be too late to save those around it.

LIFE CYCLE

With wet, cool conditions, the disease can spread rapidly, either through the soil or from countless airborne spores, which can spread up to 20 km on a windy day. The fungus can only take hold when the leaves are wet consistently during the night.

DAMAGE

Phytophthora in Latin means "plant destroyer." This disease is as serious as it gets. The fungus destroys leaves quickly, effectively turning them into lifeless brown husks.

CONTROL

This fungus gets hold of the leaves when they are wet during our cool nights. The best way to avoid the disease is to keep the leaves dry. Water early in the day (or not at all if not needed) so the leaves can dry before the cool evening falls.

In spring, purchase your seed potatoes and tomato plants from a vendor that grows them on site or imports them from a nearby source. The disease spreads by being imported from regions where it's prevalent.

If a plant does become infected, it must be destroyed very quickly. Throw it in the garbage or burn it. Do not compost an infected plant.

You can usually still save the plants around the doomed, infected plant if you act fast. Use a powder with copper sulphate as the active ingredient. Sprinkling this fungicide on the leaves will essentially deprive the fungus of the moisture it needs to do its dastardly business.

Do not use the infected patch of garden to plant potatoes or tomatoes in next year. Although the fungus can't survive our winters, it's best to be safe.

The potatoes of infected plants may be okay. Watch them carefully for brown spots and, if you usually grow your own seed potatoes, buy local potatoes to plant the next spring just in case.

Late blight can affect tomatoes (above) and potatoes (below); watch for signs during wet summers.

Potato Scab

After a long season of planting, hilling, watering and weeding, unearthing a mound of scabby potatoes at harvest time is downright infuriating. Potato scab (*Streptomyces scabies*) is a bacteria that lingers in the soil waiting for the chance to strike.

IDENTIFICATION

Scab varies in appearance. The lesions will always be round, and in extreme cases they'll coalesce into a mass. Sometimes they will have a brown, corky look; sometimes they'll be pitted into the spud; and other times they'll be raised and abrasive. It's all the same scab; the variety reflects varying environmental conditions.

LIFE CYCLE

The potato scab bacteria can survive for a long time without potatoes, feeding on trace amounts of organic matter.

It lies waiting in the soil until it can infect a potato, and enters via tiny wounds or lesions on the spud's flesh.

DAMAGE

The good news is that scab damage is purely aesthetic. The spuds are still edible, but they're ugly and you may have to peel down a little deeper to reach unblemished flesh. The worse the outbreak, the deeper the scab will penetrate.

CONTROL

The bacteria can live, and overwinter, in the soil as long as there are tasty decaying plant bits to eat. If you have scab, or even if you don't, rotate all your veggie crops annually. Don't plant other root veggies in the taters' spot, or they might catch the scab, too.

Use a pH meter to check your soil. Scab thrives in soil of 5.5 or higher, while potatoes' optimum growing range is between 5.0 and 5.5. If your soil measures above 5.3, amend the planting area with some peat moss or sprinkle spruce needles over the soil so the acid leeches down while watering.

Scab tends to grow best when the developing tubers are kept dry. When the potato plant is in flower, the tubers are at the fastest and most vulnerable point in their development. Make sure to keep the soil moist (though not overly moist, as that can cause other issues) during this period.

Some spuds are more resistant to scab than others. If it's an issue in your garden, look for Chieftan, Norland, Russet Burbank or Superior potatoes. Always check seed spuds for scabby lesions before buying, and only buy spuds from vendors you trust.

Recently, molasses has been found to contain fatty acids that will kill the pathogen causing potato scab within 48 hours. Blackstrap is the very viscous molasses remaining from sugar cane processing. Mix 250 mL unsulfured blackstrap with 20 L water; let the mixture sit for 24 hours before you apply it to let the microbes wake up. Pour it on the soil right before planting time. Reapply a few times throughout the growing season, pouring it around the hills but not on the leaves. On top of controlling scab, molasses promotes beneficial bacteria to breed in the soil and will lead to healthier dirt overall.

Both 'Chieftan' (above) and 'Kennebec' (below) are scab-resistant varieties.

Bees

Honey bee

When I was young, the sound of a still, sunny, summer day in the garden was the intoxicating buzz of bees, dancing among the flowers on their busy errands. I would follow them from bloom to bloom, watching how they scooped as much pollen onto their legs as they could before they went buzzing off again, wobbling in the air like an overloaded plane.

These days, our garden is quiet. In spring, when the apple, sour cherry and apricot trees open up into their full glory, there are precious few bees floating between the blossoms. Canada's pollinators, especially the humble bee, are in trouble thanks to habitat loss, overuse of pesticides and disease.

BEE 101

Honey bees (*Apis* spp.) are among the most valuable insects on the planet, and it has taken the disaster of so many of them dying for people to realize that they are worth fighting for. Every hive produces about 90 kg per year of one of the most perfect foods on the planet: honey. Honey is the only food that never goes bad; edible honey has been found buried deep in Egyptian tombs. Without bees, a third of the food we eat would simply disappear due to lack of pollination.

Urban beekeeping is becoming popular as cities across Canada are starting to learn how valuable bees are. A beehive pollinates crops within a 5 km radius, which means an urban hive can benefit tens of thousands of people.

The queen bee is the heart of the hive, and each queen demands loyalty from her subjects using unique pheromones. In winter, all the bees cluster around her and, boosting energy from the sugars in their honey, beat their wings so furiously that they maintain the hive's interior temperature at 35°C, even in January!

Other bees are important as well. Bumble bees (*Bombus* spp.) are excellent native pollinators. They are good-natured and are active very early in spring until well into fall. Each colony lasts only one summer, but mated females overwinter and establish new colonies in spring.

ATTRACTING THEM TO YOUR GARDEN

Bees are like birds—there aren't enough to go around, so we need to entice them. The more food we give them, the more food they give us. Without bees, our apple trees, raspberry patches and strawberry pots would be bare.

Bees prefer sunny areas that are sheltered from the wind. The middle of a yard that is surrounded by trees and houses is the ideal place for a bee-friendly flower bed.

Bumble bee

Plant your bee-friendly plants in clusters of at least three. A bee will always buzz toward a group of plants over a single one. Try to plant a variety of flowers, and make sure to include different shapes. Different species of bees are attracted to different shapes of flowers.

Luckily, bees are attracted to the same vibrant flowers that you're attracted to. They love colour, especially violets, so don't be bashful about planting flowers that steal the show.

Bees love native perennials much more than exotic annuals. Some of the best plants to attract them are black-eyed Susans, lilacs, echinacea, rhododendrons and sunflowers. Try to plant groupings of perennials that will bloom in stages throughout summer.

KEEPING THEM THERE

Pesticides kill bees and other beneficial insects as well as the pests they're meant to kill. In the case of bees, the chemicals collect on the flowers and accumulate in bees' bodies as they buzz from one blossom to the next. Never spray a pesticide on a plant that is in bloom, and if you must use chemicals, start with the least toxic (such as horticultural oil) and go from there.

COLONY COLLAPSE DISORDER

It's impossible to overstate how extraordinary bees are or how vital they are to our future as a species. Worldwide, about 200,000 species of bees pollinate 80 percent of all flowering plants. Without bees, grocery store produce sections would almost disappear, with everything from apples to beans and even coffee vanishing from the shelves.

In the past 10 years, an estimated 40 percent of U.S. honey bee colonies have disappeared owing to colony collapse disorder (CCD). Our bees are disappearing at an alarming rate, and finding out why has become a race against time. While there are many theories, recent research has paid special attention to two of them.

Plant native perennials such as asters to attract bees to your yard.

"NEONIC" PESTICIDES

For years, scientists have searched for the root cause of CCD, with theories ranging from pesticides to parasites to cell phone signals. A 2014 study out of Harvard University has found a smoking gun: a new class of neurotoxin pesticides called neonicotinoids, or neonics for short.

Neonics enter young plants' systems either via the seed being treated or a spray application; many of Canada's largest crops are planted with neonic-treated seed. Treated seed is appealing to farmers and industrial plant growers because it reduces the need for, and therefore the cost of, future pesticide treatments. However, it turns the plant into a death trap for visiting bees.

Bees exposed to neonics start to shake uncontrollably and suffer mental breakdown. The condition has been compared to an extreme form of Parkinson's and Alzheimer's combined.

As consumers, we can help by applying pressure on retailers not to sell neonic-treated bedding plants. If you don't buy your spring plants locally grown, ask if the plants have been treated with neonicotinoid pesticides. Although you'll get a lot of blank stares, awareness is growing. Some major independent garden centres in the U.S. are already promoting that they do not sell neonic-treated annuals, and in 2013, the European Union instituted a two-year ban on their use.

As gardeners, we can help by reducing our reliance on pesticides. Please consider spraying anything in your yard to be a last resort, and never spray while trees are in bloom.

VARROA MITES

The aptly named *Varroa destructor* mite is a major cause of CCD. Introduced to Canada in the late 1980s, it destroys colonies by sucking drones' blood and weakening them to the point of disease or deformity.

Varroa mites travel quickly from hive to hive and, if not treated early, tend to be fatal to the colony. Although there are effective actions that beekeepers can take if the mites are detected early, early detection is difficult.

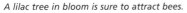
A lilac tree in bloom is sure to attract bees.

Ladybugs

Seven-spot lady beetle

The past few years have seen a lot of buzz for a pretty predator called ladybug. We've all heard that they're one of the best bugs to have in your garden, but how do you get them there and, once they've come, how do you convert opportunistic visitors into an aphid-smashing army that sets up permanent barracks?

Ladybugs are arguably the world's most beloved insect, and almost all cultures believe that killing them is inviting misfortune. Depending on where you are, if a lady beetle lands on your hand, superstition dictates that you're either getting new gloves (England), getting married within a year (Belgium, if you're a single woman), or that you can expect any ailment you have to fly away with it (France). In the 1800s, doctors were even known to jam them into cavities to alleviate toothaches or prescribe them to treat measles.

During a particularly nasty aphid infestation in the Middle Ages, Europeans prayed to the Virgin Mary to save their crops. Legions of ladybugs appeared, and they were henceforth named the "Bug of our Lady" in tribute.

LADYBUG 101

Entomologists wince at the term "ladybug." Lady beetles—their correct name—actually aren't a true bug. Of the over 5000 species of the lady beetle family (Coccinellidae), 500 of them hail from North America. They've been farmers' friends for centuries and have proved to be arguably the most useful agricultural insect in the world.

Ladybugs are so vicious that I'm glad they're on our side. They're aphid-eating machines, devouring up to 50 aphids a day and about 5000 over their lifetime. Depending on the lady beetle species, scale and other soft-bodied pests are doomed, too.

Lady beetles can have three to four life cycles in a year, with females laying dozens of eggs a day on the undersides of leaves. Emerging larvae look like tiny black caterpillars with an orange blotch. Be careful not to confuse them with bad critters.

If aphids run short, ladybugs can live on pollen and nectar for a short time, after which they will cannibalize each other before starving or seeking protein elsewhere.

Convergent lady beetle

GETTING THEM TO YOUR GARDEN

If you want a larger population of lady beetles in your garden, you can either buy them or attract them. The former option is quick and easy, but you'll

need to know a few things before opening that bag, and the latter option may require some minor garden retrofitting.

Like all critters, lady beetles have favourite plants. Adding herbs such as dill and cilantro, annuals such as sunflowers and zinnias, and perennials such as yarrow, solidago and alyssum will help lure ladybugs to your side of the fence.

A ladybug's dream garden has a wide variety of plants in it (including those listed above), which are mature enough for their leaves to overlap. It's also watered regularly enough that there is access to droplets to drink. If you keep your garden on the dry side, consider submerging a small container of water with rocks and twigs to land on.

The other crucial aspect to attracting ladybugs, which some of you may not like, is food. Aphid-free yards are also ladybug-free; the beetles will move on until they find a food source. If you want to maintain your ladybug population, you'll need to look the other way at the odd aphid (unless the infestation becomes severe) so your troops don't starve.

To get your army started, you can buy ladybugs by the hundreds at large garden centres. Call around first to make sure they carry them, and check the bag; they will have an expiration date.

There are a few tricks to releasing them to ensure they don't all scatter. Keep them refrigerated (not frozen) until you're ready to use them. The cold will slow down their metabolism so they will live longer.

Ladybug adult (above); larva (below)

Ladybugs tend to fly during sunlight hours, so wait until twilight to release them; it will give them a chance to get comfortable before the sun makes them restless. They will be thirsty, so spray water across the release area. Take the time to sprinkle them over a large (preferably aphid-infested) area; if you dump them all in one spot or just open the bag, most will waste no time flying away.

KEEPING THEM THERE

When a gardener sees aphids, the urge to kill is almost instinctual. Every day I see people come into the greenhouse looking for the most toxic chemical on the shelf to douse their yard with. While this is the fastest way to kill aphids, it's also a trap.

Once you establish a viable lady beetle population (i.e., see them hanging out on plants throughout the growing season), resist the urge to spray pesticides, or you will have to start all over again. Spraying pesticides is chemical warfare and wipes out all insects, good and bad. The aphids will re-establish their population much

faster than the ladybugs will, and before long, they will be worse than before.

It takes a long time to establish a viable ladybug army, but it's priceless once you have it. When aphids strike, trust that your resident army will deal with it. It won't be overnight, and they may need to bring in reinforcements, but they will win.

Eventually, your yard will become a balanced ecosystem wherein predator and prey endlessly skirmish but neither side wins. Ladybugs, not wanting their offspring to starve, will lay eggs only on aphid-infested plants.

Take care when cleaning up your perennial beds in fall. Ladybugs live up to three years and overwinter among old leaves near the soil, at the base of a tree or under rocks. Dozens or hundreds huddle together, and if you find one of these spots, make sure to leave it unmolested. Unfortunately, slugs also love these spots, so make sure to check for their caviar-looking egg clusters before the snow flies.

Ladybugs love dill and sunflowers; including certain plants in your garden will encourage the beetles to take up permanent residence.

Spiders

Goldenrod spider

If our gardens were Hollywood movies, every good bug would be as cute as a ladybug and every bad bug would be as ugly as a spider. Spiders (Araneae) are one of the good guys, but unfortunately for them, they look like villains and are often persecuted as such.

If we knew what spiders really did all day, we wouldn't expect them to look pretty. They are the down and dirty foot soldiers of our gardens, eating anything they can get their eight legs on with alarming ferocity.

People seem to be either afraid of them or fascinated by them (count me proudly in the latter group). Spiders aren't aggressive and will bite only if threatened or squeezed. If you're currently among the multitude who kill any spiders they find, I truly hope that, after reading this, you'll live and let live. When you find one in your home, please collect it carefully in a glass and give it a fresh start outside.

Wolf spider

SPIDER 101

While many people believe that the majority of spiders spin webs, the opposite is actually true. Webs are very visible, so orb-weavers get all the press, but our most common spiders are the solitary hunters. Wolf spiders (Lycosidae), jumping spiders (Salticidae) and goldenrod spiders (*Misumena vatia*)—the yellow or white ones lurking in flowers—account for the eight-legged majority in Canadian gardens. Flip over some mulch or look under stones, and there's a good chance you'll see a furry little brown wolf spider scurrying away.

Spiders are true carnivores and will eat almost anything they can catch. Depending on the species of spider, this can include garden pests such as aphids, mosquitoes (in great numbers), flies, caterpillars and pretty much anything else that gets snared in their web or wanders a little too close.

JEWEL SPIDERS

If I asked you to describe the biggest, scariest-looking spider around, I'd bet my bottom dollar that it would be a jewel spider (*Araneus gemmoides*), especially if you hail from the Prairie provinces. Also called cat-faced spiders, they are orb-weavers that appear in late summer and spin large, circular webs

Jewel spider

If you have a jewel spider sharing your space, consider yourself lucky.

overnight, and often in the worst places (like across your front door).

Jewel spiders are one of Canada's largest, with females often 2 cm across. Up close, they are straight out of a horror movie. They are also one of the best critters to have in your yard, and if people knew more about them they would be trying to entice them onto their deck instead of chasing them off.

They're harmless to humans, biting only if their life is threatened. Even on those rare occasions, their bite is very mild. They aren't so passive to mosquitoes, however, and on high skeeter years, their webs spring up everywhere. Celebrate every web you see, because jewel spiders can eat their own weight in blood-suckers every day.

POISONOUS SPIDERS

We're lucky in Canada to have very few poisonous spiders. The two talked about are the brown recluse (fiddleback) spider and the black widow spider. Bites from both of these spiders are serious but very rare.

The brown recluse (*Loxosceles reclusa*) isn't native to Canada, but the odd bite is reported when one has hitch-hiked in from the warmer U.S. They aren't something to be frightened about, and chances are that you've never seen one, but just to be safe, don't make a habit of handling unknown spiders.

Black widows (*Latrodectus hesperus*) are generally found in southern regions of Canada. They're shiny, black, and females have that tell-tale red hourglass; you'll know it when you see it. If you see one those spiders inside your home, kill or get rid of it immediately (in a safe manner). If you see a number of them, call an exterminator.

Black widow

Earthworms

Of the humble earthworm (*Lumbricus* spp.), Charles Darwin said it best: "It may be doubted whether there are many other animals which have played so important a part in the history of the world, as have these lowly organized creatures."

Native to Europe but introduced worldwide, earthworms are vital to the health of our garden soil. A square acre of healthy earth has about 50,000 worms working hard to keep it that way, and there are things that we can do to ensure that they flourish in our home gardens.

Worms have become big business in recent years, with more people buying worms to set up worm-composting (vermicomposting) systems in their homes or just looking for worm casting–enriched organic fertilizers.

Having earthworms in your garden soil leads to healthier plants.

EARTHWORM 101

Nicknamed "nature's plow," worms are the humble workhorses of the garden. Their tube-like bodies tunnel through soil, soft or hard, loosening it as they go. Plants' delicate root systems thrive in well-aerated soils where they can expand freely into air pockets, leading to healthier plants and better harvests.

Besides loosening compacted soil, worms leave castings behind that are so stuffed full of nutrients they have become a valued ingredient in many organic fertilizers. Worm castings contain about seven times the nutrients as common compost, including micronutrients that plants can't get anywhere else.

The castings act like sponges, soaking up water and retaining it far longer than the surrounding soil. Having ample worm castings increases overall water retention in the soil and, in doing so, will help plants better tolerate droughts.

Bacteria living inside worms' bodies have the ability to actually break down some toxic chemicals that accumulate in garden soil. Their bodies actually detoxify some fungicides and pesticides, though it's often at significant cost to the worms' health.

NURTURING YOUR WORMS

Having a healthy worm population can take your veggie garden from fair to flourishing. It's easy to make your garden inviting to worms if you think like a worm.

Worms love moist soil that's amply loaded with yummy rotting organic matter. Let an area 1 square metre in the garden accumulate leaf litter and, if you're keen, bury some kitchen scraps a spade depth below the surface. Worms don't travel far, so make sure the plot is adjacent to the garden you want them to move into.

Keep your garden consistently watered; parched soil is dangerous to their moist bodies. Try to keep digging, especially rototilling, to a bare minimum. Once you've established your squiggly population, protect it by not spraying harsh chemicals in the garden.

You can also buy worms for the garden, just as you would for your tackle box, but you really don't need to. If you keep a worm-friendly garden, you'll get them naturally. If your garden isn't worm friendly, they'll die or leave no matter how many you buy.

EARTHWORMS INVASIVE IN FORESTS

While they're beneficial in the garden, it's important to keep them in the garden and out of our forests. Earthworms damage Canadian forests by breaking down leaf litter, which many other species of plants need to survive, faster than usual. Worm-invaded forests lose a lot of biodiversity at the forest floor level.

All of Canada's native earthworms were wiped out during the last Ice Age, and our forests evolved to flourish without them. Now that we're widely importing earthworms from Europe, they're wriggling their way back into forests that don't need them.

The worms in your garden are not going to ravage the nearby forests. Left alone, they will rarely travel more than 10–20 m in a year, and they rarely pass artificial barriers such as roads.

If you use them in your tackle box, however, then you're giving them a free ride into pristine forests. Earthworms have invaded our forests as a result of fishing, not gardening. Anglers, please never throw your unused wriggling bait into the forest. You're actually aiding and abetting an invasive species.

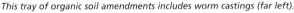

This tray of organic soil amendments includes worm castings (far left).

Nematodes

The proliferation of beneficial nematodes is an exciting development in pest control. They're 100 percent safe to humans, pets, plants and beneficial insects, and they're ruthlessly effective. Most people still haven't heard of garden nematodes, but when they do, we're going to be seeing a lot more of them.

Nematodes, which are also called roundworms, are a massive group of tiny worms (around 1 mm long or less). When I say massive, what I really mean is staggering. An estimated 1 million species live in every corner of the world, from the polar regions to the deep seas. They've even been found in mine shafts almost 4 km below the surface of the earth.

It's impossible to imagine just how populous nematodes are. In some areas there are an estimated 1 million individuals per square metre of earth. They account for 90 percent of all life on the ocean floor and about 80 percent of all individual animals on earth. Give that a second to sink in.

NEMATODE 101

About half of the 25,000 described species of nematode are parasitic, meaning that there's a nematode invader for every species of host (including us and our pets). The species most commonly used in gardens are harmless to everything but their intended prey; they are assassins trained only on their intended victim.

Nematodes in the soil enter the larvae of their prey via their mouth or another opening and begin excreting certain bacteria. The bacteria convert the victims' internal tissues into food for young nematodes, usually causing the death of the host within 48 hours. Several nematode generations will feed on the insect before migrating into the soil to seek a new host. When they run out of victims, the nematodes starve and biodegrade.

Beneficial nematodes kill over 200 species of garden pest, specializing in soil-dwelling grubs such as the sod webworm. They also devour flea beetles, fungus gnats, root maggots, weevils, Japanese beetles, leaf miners, ticks, thrips… The list goes on.

USING THEM

Call your nearest large garden centre to see if they carry nematodes. If they don't, tell them they should and go to a reputable dealer online. They are becoming easier to find every year as the word spreads.

Nematodes are sold dried and wake up with contact with water. Following the directions on the package, mix them in a watering can, backpack sprayer or the like and apply to the affected areas of your yard. Keep mixing the water to make sure the critters don't sink to the bottom.

Make sure the ground is slightly moist when you're applying nematodes so they have something to drink and so it's porous enough for them to wriggle downward. The air temperature should be between 10°C and 30°C. Try to apply them in the evening, early morning or on a cloudy day, as direct exposure to harsh sunlight can be lethal.

Although they can stay viable for weeks when stored cold, apply them immediately upon buying and always check the expiry date first. Avoid fertilizing the lawn for two weeks before and after nematode application, especially with high-nitrogen blends (available in spring). The nitrogen will burn their delicate bodies. Avoid applying herbicides, as well.

They set to business immediately upon hitting the soil, deploying like shock troops in search of victims. For specific outbreaks, you may need to reapply them in two weeks to nix the pests' life cycle. You'll probably see dead insects lying around, but the nematodes will be too small to spot.

Depending on where you live in Canada, winter will wipe out either all or almost all of your nematode army. It's best to reapply them in spring, because even if a few survive, they won't be enough to successfully battle the maggoty hoards.

Introducing certain nematodes to your vegetable garden and lawn will keep pests under control.

Canada Thistle

If you're a gardener, I'll bet my bottom dollar that just reading the name "thistle" conjures memories of scratched hands and sweaty foreheads as you try to pull it from its tap-rooted moorings. Canada thistle (*Cirsium arvense*), also called creeping thistle, is tough, aggravating, ubiquitous and here to stay.

Canada gets a bad rap on this one. Although most of North America calls it "Canada thistle," it was actually imported from Europe and Asia in the early 1600s. It's invasive there too, where it's also affectionately referred to as lettuce-from-hell and bull thistle.

Its roots and peeled stem are edible, albeit bland, I'm told. The roots are the most nutritious, though their legendary tendency for promoting flatulence doesn't exactly make them a date-night favourite.

On the plus side, thistle foliage is an important food source for the remarkable painted lady butterfly. First Nations used infusions from the roots to treat mouth infections and chewed the leaves to soothe toothaches and sore throats.

IDENTIFICATION AND PROPAGATION

Thistles are fast growing, with a thick central stalk and dark green, oblong leaves all covered with spines. They can reach almost 2 m in height and, by the time they get that large, their stem is so fibrously thick that it will take a hatchet to bring them down.

In early to midsummer, they bloom small, tufted, soft purple flowers. If these pretty blossoms don't strike fear into your heart, they should. Each plant produces about 1500 seeds, which will leap into the summer wind and take up residence around your yard. The seeds can stay viable for over a decade, waiting for the right conditions. No flowers means no seeds, so don't let them flower.

Thistles also spread quickly on horizontal rhizomes that stay close to the surface. In one season, a sprouted seed can colonize an area 2 m in diameter with rhizomes, from which dozens of clones will spring up at will.

It's a bully once established, quickly forming dense stands resembling spiny miniature forests, choking out everything else.

CONTROL

First off, if you have thistles in your yard, don't expect to ever truly be rid of them. After all, if noxious weeds were easy to eradicate, they wouldn't be noxious anymore.

Thistles' tap roots are epic. Pull young plants out quickly before their roots dig into the earth like a drill. If the plant dies or gets yanked, all that's needed is 3 mm of root to start growing a new one.

Even though it will keep producing new plants from root leftovers, the most effective control is still good old-fashioned yanking. Consistent pulling will, eventually, deprive the root and rhizomes of life-giving photosynthesis. Stress the plant enough, and over the years, the rhizome will send up fewer clones. And, as mentioned, pulling them before they flower will stop them spreading by seed.

Painted lady butterfly

Creeping Bellflower

I recently read Max Brooks' page-turning undead romp, *World War Z*. The book is a meditation of what a global zombie plague, and our subsequent reaction to it, would look like. In the book, infected victims don't become zombies right away. It takes days, time enough for them to mingle with you and become accepted as just another human. That is, until they eat you.

Zombies spread inexorably because as long as one survives, its sole determination is to spread and begin the plague anew. Their absolute focus on creating more of themselves is what makes them so hard to eradicate and enables them to spread like wildfire.

As I was reading, the zombie behaviour smacked as oddly familiar. The other day, a customer came in with a sealed grocery bag and told me that it contained a volunteer plant in her garden that seemed to be spreading quite quickly. She said that its pretty blue flowers had made it hard to pull in previous years, but that this year they seemed to be sprouting up everywhere. I didn't have to open the bag to know that she had a zombie weed outbreak.

Bellflowers aren't native to Canada and were introduced from Europe as an ornamental plant. If you search online, you will astonishingly find the creeping species for sale, and some sites contain posts from people who rave about buying or finding this blue treasure. Most of the reviews, however, are warnings.

IDENTIFICATION AND PROPAGATION

Creeping bellflower (*Campanula rapunculoides*) produces clear blue, trumpet-shaped flowers along a tall, stately stem. Upon seeing its spontaneous blooms in your perennial bed for the first time, your reaction will probably be, "I didn't plant that, but it's pretty enough to leave alone."

For the first couple of years, it will be pretty and deceivingly well-behaved. But 20 cm beneath the surface, it's building a rhizomal substructure that is shuffling under flower beds and lawn alike.

By the time you notice it's a problem, zombies are sprouting all over your yard. The heart-shaped leaves appear en masse in perennial beds and lawns and quickly choke out any resident plants.

Creeping bellflower thrives in dry or wet soils, full sun or full shade. It can lie dormant for years and, if there are no insects to pollinate it, it will pollinate itself to make seeds.

It spreads by both rhizome and seed, and any shred of rhizome is enough to create a new, single-minded army of weeds. Each plant can produce 3000 seeds, and each seed comes equipped with wings for drifting across fences to quickly plague entire blocks and neighbourhoods.

CONTROL

Most importantly, if you see a pretty blue flower appear unexpectedly, yank it out. There are several other *Campanula* species that boast the same clear blue flowers and won't drive you to distraction.

You can slow the spread by pulling them before they bloom, which will stop the spread of seeds and will start to deprive the rhizome of the photosynthesized nutrients sustaining it. The rhizomes run so deep that you would

Don't be deceived by its pretty flowers; creeping bellflower is a noxious, hard-to-get-rid-of weed.

have to excavate almost 30 cm of earth to reach them, and even so, if there is even one shred left, it will create a new batch of zombies. It will take time, but pulling every flower you see will, over the years, severely weaken the rhizome.

Don't bother spraying Kill-Ex on it. Creeping bellflower is immune to 2,4-D (the active ingredient). Round-Up, containing glyphosate, will slow it down but, in the process, will kill everything green it touches and, yes, the zombies will keep coming.

Always check the ingredients on wildflower seed packages, as it's been known to find its way into the mix. Never buy wildflower seed packs that don't list all the species inside.

Creeping bellflower has recently been listed as a noxious weed across Canada, and bylaw officers have and will be out in force, issuing thousands of citations to clean up infested yards. If cited, homeowners have 10 days to clean up, or a contractor will do it to the tune of a few brown bills.

Peach-leaved bellflower is a related, but well-behaved species welcome in your garden.

Field Bindweed

If you love the annual vine morning glory, then this villain of the garden will look uncannily familiar. Like its coveted ornamental cousin, field bindweed (*Convolvulus arvensis*) yields broad, beautiful flowers in midsummer. Unlike morning glory, bindweed's astonishing speed of growth and ruthless habit of choking out anything in its way has made it one of Canada's most loathed garden weeds.

Bindweed creeps quickly and twines around anything and everything in its path, eventually creating a thick carpet and devouring all sunlight. As bad as it is for gardeners, it's potentially devastating in farmers' fields.

Introduced from Asia and Europe in the 1800s, it is now found in almost every province (Newfoundland and P.E.I. are off the hook). It weaves deep nets of rhizomes underground and a dense carpet of leaves above-ground to colonize and conquer any sunny spot it encounters. If ignored, it will consume your yard.

IDENTIFICATION AND PROPAGATION

Bindweed is easiest to spot when it is blooming, but try not to let it get that big. It starts low to the ground and prefers open, sunny areas and nitrogen-rich soils.

Look for triangular leaves that always point outward. It will have a creeping habit, and by the time you find it, it may already be slinking around your ornamental plants.

If you or your neighbour (or the guy 15 doors down) has it and a bird "deposits" a seed in your yard, it can establish itself astonishingly fast. In one growing season, a seed can spread rhizomes 3 m out and send up dozens of growing shoots.

In midsummer, larger plants will bloom and, yes, they will be beautiful. The round, funnel-shaped, white flowers have a soft texture and delicate fragrance. Don't be fooled; the seeds are growing inside.

Each plant can produce up to 500 seeds, which birds promptly eat and spread across the garden. As the seeds stay viable for 30 years, once they're spread, you'll be fighting this weed for a long time.

CONTROL

Like many of Canada's worst weeds, field bindweed has adapted multiple ways to spread and make your life frustrating. If you see it blooming, don't enjoy those pretty flowers for too long or it will go to seed.

Bindweed rhizomes grow deep and, like quack grass (see page 155) and creeping bellflower (see page 150), every time you tear a rhizome, it creates a new growing end. For this reason, don't bother digging. Tear the plant off at soil level as soon as you see it. If it can't photosynthesize, the rhizome will eventually starve. Be persistent; it takes a while, but it works.

Spreading organic mulch, such as cedar chips, throughout your perennial beds will substantially reduce bindweed outbreaks (and other weeds, as well). Bindweed hates the shade, and that on top of needing to push through mulch will stop a lot of shoots.

As a last resort, a broad-leafed herbicide (such as one with the active ingredient 2,4-D) will slow down the weed, though it's very rare that it will stop it completely given that a mature plant's rhizomes often dig 6 m into the earth. Treat it in fall, and never exceed the recommended concentration amount on the bottle.

Dwarf morning glory is a very pretty, non-invasive cousin of field bindweed.

Quack Grass

Most of the entries in this book afford me the happy task of reassuring people that intimidating gardening tasks are easier than they think. Unfortunately, it's my job to tell you that quack grass is just as bad as you think it is.

Known also as couch grass, twitch grass and, most aptly, devil's grass, quack grass (*Agropyron repens*) is a scourge of both farms and gardens in every part of Canada. It's listed as one of our top three most vile weeds and is estimated to infect over 50 percent of farmland in the country.

Quack grass's Latin name translates to "sudden field of fire," in rueful recognition of its ability to quickly infest large swaths of earth. Native to Europe and western Asia, it's thought to have caught a ride across the Atlantic Ocean centuries ago with cereal crops.

Quack grass rhizomes grow rapidly, sending up new shoots.

IDENTIFICATION AND PROPAGATION

When it first sprouts, quack grass will look like regular grass. However, quack grass grows at a rapid rate, far outpacing the lawn, and it won't take long to develop the broad, coarse leaves that are perfect for making the most wonderfully irritating duck sound in the world.

Quack grass spreads by both seed and rhizome. Each mature plant stem will yield about 25 seeds, which stay menacingly viable for years once on the ground.

The white, fleshy rhizomes, somewhat resembling sprouts, grow rapidly until they establish a thick mat a few inches under the soil surface.

When I say they grow rapidly, I'm not kidding. In the cool spring months, when it's most active, its rhizomes can sprint up to 2.5 cm per day. New grass blades can emerge from any point on the rhizome.

The rhizomes grow from any and every tip. If you break one in half, it suddenly has two new tips to grow and spread from. The tips are strong enough to push through potato tubers and, over time, even asphalt.

CONTROL

It's a common mistake to rototill the rhizomes into little pieces. While this method may work on some other weeds, quack grass is a super-villain weed, and each shredded bit of rhizome will grow into a whole new network of frustration.

If there are no other plants in the infested area, all you'll need is a little determination. When the soil is moist, carefully dig up the soil and remove the white, fleshy rhizomes by hand. Take care not to break them.

Once the rhizomes have been removed, lay down a thick layer of wood mulch, with an optional layer of cardboard underneath, to smother any new growth. Keep a close eye around the edges of the mulch, as any missed rhizomes will creep—quickly—to where they sense sunlight.

It's important to pull new green growth as soon as possible. The rhizomes feed on nutrients derived from photosynthesis, and if deprived of those nutrients, they will eventually starve. On the other hand, a patch of quack grass left unseen and un-pulled will strengthen the rhizomes, and you also risk it going to seed and spreading that way.

If quack grass has infested your lawn, shrubs or perennial beds, settle in for a war of attrition. The rhizomes wrap around other plants' roots, making wide-scale, effective excavation impossible. Stay diligent and keep pulling. Laying a thick layer of wood mulch around your existing plants will slow down emerging blades, but take care that it's not so thick around your plants' stems that they rot from never drying out.

Because quack grass isn't a broadleaf weed, it's immune to selective herbicides such as Killex, which are designed not to kill lawns. If you don't want to kill all your other plants as well, your only chemical option is to tediously paint a glyphosate herbicide such as Round-Up on the blades themselves. Do it on an afternoon hot enough that the blades will pull the poison thirstily into their rhizomes. Don't use Round-Up if you have pets or children using your yard.

If you have a quack grass infestation, be prepared for a long battle to get rid of it (above); try to pull quack grass before it goes to seed (below).

Chickweed

One of the most ubiquitous weeds in Canada, chickweed (*Stellaria media*) is fast to grow, hard to kill and surprisingly good in salads. It grows into a dense, tangled mat and spreads quickly via surface nodes and rapidly developing seeds.

While a significant nuisance in yards, chickweed is a menace in fields. It has a particularly sadistic fondness for choking out cereal crops and can reduce barley yields by a staggering 80 percent. Needless to say, it's in the cross-hairs of gardeners and governments alike.

IDENTIFICATION AND PROPAGATION

Imagine what a thick, tangled, curly-haired, green wig would look like if you threw it on the dirt in a heap. That's basically chickweed. Its unique habit is to grow in and around itself, getting deliberately tangled, so the only way to pull it out is by the fistful.

While it probably won't be on your radar during drought years, chickweed thrives during cool, wet growing seasons. It grows quickly, with a germinated seed sprinting to flower in as little as a month.

The Latin name means "in the midst of little stars," which is exactly what the flowers look like. Hard to see if you're not looking for them, their petals are split, making five look like 10. Don't let their delicacy fool you; an army of seeds lurks within.

In most weeds, flowering is a red-alert warning to yank it before it goes to seed. Chickweed has devilishly developed the ability to set seeds immediately after it flowers, so by the time you spot the little white blooms, about 800 seeds are already massed and ready to attack.

CONTROL

As with most nasty weeds, the best control is the ol' grab-and-pull. Don't be frustrated when you don't get the roots. Flimsy stems combined with shallow roots make it almost impossible to pluck up the whole package.

Persistent weeding is important, but doing it before it flowers is crucial. Walk through the garden often and keep your eyes to the ground; it grows so low it's easy to miss.

Herbicides are a last resort and should only be used if the problem is getting out of control. Both 2,4-D and glyphosate are effective controls. Never apply herbicides on chickweed you plan to eat later.

OTHER USES

Yep, you read that right: chickweed has uses other than just being annoying. Its fresh leaves (the younger the better) are chock full of vitamin C, magnesium and other nutrients. You can even buy chickweed herbal supplements in health food stores.

It has a subtle, somewhat grassy taste and is surprisingly good in salads and soups, though if using it in soup, only add it in the last five minutes or it will basically disintegrate. You can use the stems and flowers, as well.

Chickweed is a magnet for other insects to come into the garden. This is a double edged sword, for while it attracts ladybugs and other beneficial critters, pests such as thrips come, too.

It has anecdotal medicinal uses, though there's precious little medical research to back up the claims. People have eaten it for stomach and lung problems, rubbed it on their skin to treat boils and abscesses, and even used it to make a diaper rash cream. Just for the record, I don't endorse its effectiveness for any of that.

Try to pull chickweed before it flowers, and be persistent.

Dandelion

As adults, we indoctrinate ourselves to think of dandelions as the faceless enemy that will lay waste to our lawns and quickly forget how much we loved them as children. Besides rubbing the yellow colour onto themselves, blowing seeds, popping off heads and making bouquets are all quintessential parts of childhood. If children play in your yard, please think twice before reaching for the chemicals.

However, as pretty as they are, dandelions (*Taraxacum officinale*) are also a famously invasive and persistent weed. For those who covet the perfect lawn, getting rid of them can become the summer's white whale. Fortunately, there are a surprising number of ways to not only control them, but also appreciate them.

IDENTIFICATION AND PROPAGATION

Dandelions are universally recognized by their yellow flowers. While the spring bloom is largest, they will often bloom again in fall once the days are less than 12 hours long.

They are perennial weeds, and like the rest of your perennials, they get larger and more vigorous as the years roll by.

Their famous tap roots can grow to the size of carrots. They grow best in sparse, sunny lawns, and their leaves often lay frustratingly flat enough to evade lawn-mower blades.

The bad news is that even if your yard is currently dandelion free, it's not going to be for long. Not only can the wind carry seeds long distances, but also, those seeds remain viable for years awaiting the right conditions to germinate.

CONTROL

With thousands of seeds floating into your yard every summer, the only way to really win the war on dandelions is to prevent them from germinating. A lush, healthy lawn will choke out young dandelions before they can grow large enough to effectively compete for resources.

The goal is to deny dandelion seeds a space to sprout. Leaving the grass clippings in your lawn will help keep seeds from reaching soil level. Mowing your lawn a little higher (about 7.5 cm) will ensure that germinating seeds don't get the light they need.

Top-seed any bare spots in the lawn promptly to keep it strong and full. An annual aeration, especially with mature lawns, will keep grass vigorous enough to defend its own turf.

Dandelions are not invincible. While their sheer spring numbers tend to shock-and-awe lawn lovers into a state of panic, they are actually at their weakest right after blooming. All their reserves have gone into their flowers, and that's the best time to control them.

Pulling dandelions is still the best way to kill them, but the majority of the root has to come out with the plant to ensure it doesn't re-grow. If you pull them out of dry earth you'll get only greens, so wait for a good soaker rain to loosen the soil; the deeper it's wet, the better. If the forecast is for nothing but sun, water the area first.

If you have a leaf blower, it probably has a switch that turns it from "blow"

Dandelions are actually at their weakest right after flowering.

to "suck." Watch for when the seeds are just about mature enough to catch the breeze and strap on the seed-sucker. Simply wave the sucking end across the seed heads, and it will strip them bare. Be careful not to spill the giant mass of seeds when you empty the bag.

There are chemical solutions a-plenty, from bars you drag across the lawn to broadleaf herbicide contact sprays to dousing the entire yard with pesticides. These chemical treatments tend to be very effective, but they should be a last resort and aren't necessary for minor breakouts.

ACCEPTANCE

If you can't beat them, eat them! Not only are dandelions edible, but they are also incredibly good for you. They are packed with more vitamin A and iron than spinach, and boast more beta-carotene than carrots. The leaves are excellent for the liver and kidneys, though people with irritable bowel syndrome shouldn't partake.

The leaves are most flavourful before flowering in spring or after the first fall frost. Picked at those times, they can be added raw to salads and have a crisp taste similar to endive.

Mature, post-flowering leaves acquire an unpalatably bitter taste and fuzzy texture. A long boil or steam will tenderize them.

Never eat dandelions unless you're sure they haven't been sprayed with herbicides.

Dandelions have the frustrating ability to adapt their growth to heights below the lawn-mower blades (above); young dandelion greens are nutritious and delicious (below).

Purslane

Originally hailing from southern Europe, purslane (*Portulaca oleracea*) is now abundant on every continent except Antarctica. Unless you live in the arctic, I can pretty much guarantee you've got purslane growing nearby. Also commonly known as portulaca, moss rose or pigweed, purslane is the only weed I know of with the dual designations of noxious weed and superfood. It's as hard to get rid of as it is delicious.

IDENTIFICATION AND PROPAGATION

Purslane is a creeping annual with fleshy, succulent leaves and small but showy, yellow flowers. Being a succulent plant, it stores water in its fleshy leaves and is well-adapted to droughts. It flowers whenever it gets enough moisture to do so, so the yellow blooms can appear at any time. After flowering, seed pods form where the branches intersect and burst open when the approximately 10,000 seeds are mature.

Purslane is ubiquitous in urban areas because its drought tolerance allows it to thrive in poor, sandy soils. If an area is disturbed by being dug up, stomped on or otherwise abused so much that other plants won't grow, expect to see the oval leaves and reddish stems of purslane begin to creep.

Pull purslane before it flowers and sets seed.

Look for it at the edge of lawns along sidewalks, where the foot traffic, overflowing from the concrete, has bashed the lawn into smithereens. It's also the first plant that tends to peek through cracks in your driveway, and if not pulled, it can eventually damage the concrete.

It's a heat lover and needs a surface soil temperature over 24°C to germinate and grow. Once established, its deep taproot and the secondary stems digging in from its creeping branches make it very tough to pull.

It will grow like crazy as long as the days stay warm and long. The stems can grow to 45 cm long, until it forms a dense mat on the ground.

CONTROL

It may be delicious, but it's still a noxious weed, and eradicating it can be a pain. Try to get it when it's young and hasn't had time to build a complex root system. The only effective way is to pull it, and it's a real tug of war if you wait until it is large.

Get as much plant as possible, as purslane can rebuild itself from any leftover leaves, roots or other bits. Make sure to pull it before it has the chance to flower and produce seeds.

If it gets out of hand during a particularly hot summer, take heart that if the next summer is cooler, you'll have less of an issue.

SUPERFOOD

Purslane is a nutritional powerhouse. It boasts more omega-3 fatty acids (the good stuff in fish) than any other leafy plant. It's packed with vitamins A, B, C and E as well as calcium, potassium and more.

Purslane's mild taste has been described as being something between cucumbers and green beans. The tips are the most tender, and the leaves have a tangier taste if harvested in the morning when more malic acid has built up. All parts are edible.

Cultures around the world use purslane in everything from seedcakes (Australia) to traditional medicine (China). You can eat it raw in salads or cook it into stir-fries and other dishes as you would other leafy vegetables.

If you're going to harvest it, make sure the area hasn't been sprayed with chemicals, and look for decent-sized plants—the ones peeking up through my driveway aren't very appetizing.

In recent years, purslane has been popping up as a trendy veggie everywhere from farmers' markets to high-end restaurants. There's certainly enough of it, and I expect to see much more of it on plates in the years ahead.

Moss rose is a closely related, well-behaved cousin (above); purslane will establish itself even between paving stones (below).

Cats

I admit that I'm biased: I love cats. I love them for their beauty, for the rare moments when they show you their love, and for the fact that they've never been domesticated. The realization that my chubby, grey barn cat with less than ideal bloodlines is wild at heart is, to me anyway, part of his charm.

That being said, when cats are allowed to roam free, they can wreak havoc on neighbourhood gardens. Rather than punishing them for their instinctual behaviour, there are plenty of harmless, passive steps you can take to deter rather than attack.

PROBLEMS

Cats love to dig, and when they dig, they usually "go." Their urine is very high in nitrogen, which can burn certain plants' sensitive root systems. If you smell cat urine in the garden, water it well, and as soon as possible, to dilute the nitrogen.

Besides urine, kitty feces can be a nasty surprise when you're digging in the soil. It's aesthetically unappealing and contains pathogens that should be avoided.

DETERRENTS

When a cat does its business in your garden, it's not doing it to tick you off. It's their instinct, and no animal deserves to be harassed, chased, sprayed or worse because of that. Imagine if someone sprayed you with a hose when you tried to go to the bathroom. There are plenty of passive ways to deter cats from your patch of soil. Rather than actively striking out at the cat, passive deterrents add an element to the area that the cat will want to avoid.

No matter how tough their exterior, all cats are princesses at heart and prefer soft surfaces for their tender feet. Scatter some twigs over problem areas, a couple of inches apart, or, for serious issues, consider laying a perimeter of chicken wire around the area.

Passive cat deterrents need not be ugly. Pine cones, inserted halfway into the soil, can create attractive ornamentation while keeping cats at bay.

Cats dislike citrus, so throw old orange and lemon rinds into the garden instead of the compost. As the peels break down, they'll not only keep cats away, but also provide a small amount of fertilizer as well as valuable worm food.

Human hair, as odd as it sounds, often keeps cats away, so consider emptying your hair brushes into beds you don't want used as litter boxes. Steer clear of toxic moth balls, however.

Last but possibly most important: cats "go" where they or other cats have "gone" before, and they find that spot via scent. Washing the affected area with a hose (over and above the usual watering) will go a long way to dilute the smell and, in doing so, take away the welcome mat.

Speaking of welcome mats, some people opt to deter cats from one area by luring them into other areas of the yard. Cats love the smell of mint, honeysuckle and, of course, catnip. Planting a few of these around a patch of loose, sandy, dry soil is sure to entice kitty away from your prize perennials. A word of caution, however; this strategy may end up luring other cats into your yard that may not have ventured there before.

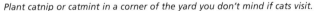

Plant catnip or catmint in a corner of the yard you don't mind if cats visit.

Rabbits and Hares

Eastern cottontail

We all know rabbits are cute. But everyone who has ever had their hard-fought spring vegetable seedlings mowed down in a single night would remind you that cute can be destructive, too.

At any given point, numerous Canadian cities are inflicted with rampant populations of pet rabbits that, either via escape or abandonment, have bred like—well, you know—and gone wild. When outbreaks happen, it usually causes a fair amount of civil discord as citizens grapple over how to deal with such cute, cuddly and destructive creatures.

IDENTIFICATION

Canada has two native species of rabbits and three species of hare. For rabbits, the eastern cottontail (*Sylvilagus floridanus*) is a little fella, about 1.2 kg. It ranges from southern Ontario and Quebec to Manitoba. The even smaller Nuttall's cottontail (*S. nuttalli*) lives in southern, arid areas of the three western provinces.

Hares are larger and considerably less cute, with lanky legs and longer ears. The snowshoe hare (*Lepus americanus*) is found in the vast forests of central and northern Canada to the tree line. The common white-tailed jackrabbit (*L. townsendii*) lives on the prairies and has acclimatized masterfully to urban environments. Southern Ontario boasts a European hare (*L. europaeus*), which was introduced decades ago.

LIFE CYCLE

Rabbits have fairly short lives, with hares averaging about five years and cottontails two years (assuming nothing eats them before that). They make up for this short lifespan by having two to six litters per year, with two to eight hungry new varmints in each litter.

In the wild, where predators are plentiful and rabbits are a favourite prey, the rapid reproduction rate is an evolutionarily evolved survival mechanism for ensuring species preservation. In the suburbs, with few predators, it can lead to wide-scale vegetable garden disaster in no time flat.

Snowshoe hare in summer (above) and winter (below)

DAMAGE

For such small creatures, rabbits are capable of staggering damage. They generally hide during the day, so most of their damage happens while we sleep. A jackrabbit can eat a pound of fresh greenery daily. Look for mowed seedlings and the hares' ubiquitous oval droppings.

While rabbits typically eat grass, freshly sprouted spring veggies and

White-tailed jackrabbit

flowers are a delicacy they look forward to all winter. If there's ample food, they will set up camp. Winter makes them turn to survival food, namely tender bark on your young fruit trees, which they will chew up to a height of 75 cm. Rabbits can easily kill a tree if they chew deep enough or entirely girdle the trunk.

CONTROL

When it comes to keeping the varmints out, there are both physical and chemical barriers to choose from. The most effective solution is to erect a chicken wire fence, 60 cm high above ground and extending 15 cm below ground (those critters can dig), all the way around the garden. If you have jackrabbits in your area, consider increasing the fence height to 90 cm.

If this potential eyesore is a bit much, try some less-visible physical repellents. Wrap tree guards around potential victim trees over winter, or, if your budget allows, consider laying low-voltage electric wires around the garden. Electric fences have the benefit of being removable for winter, but also the downside of potentially shocking the cat, dog or even your child (which is why low-voltage is so important).

Chemical repellents are effective for short periods of time, but they need to be re-applied after a good rain or each time the sprinkler comes on. Try capsaicin (hot pepper), moth balls or blood meal around your most treasured veggies; just don't spray anything on the edible portions. There are also a number of other options available at garden centres.

Freshly sprouted, spring veggies are a treat rabbits can't resist.

Squirrels

Red squirrel

I can't think of any garden varmint that elicits such extreme mixed feelings as squirrels. Kids love them, songbirds hate them, and almost every yard has these cute and destructive critters.

IDENTIFICATION

Canadian squirrels are a surprisingly diverse group. Of the 262 species world-wide, 22 occur in Canada. Of these, six are the tree species that yell at us while we walk in our yards (including two flying squirrels). The others are ground squirrels, chipmunks, marmots and a prairie dog.

The colour of squirrel you see daily will depend on where you live. If you live in Vancouver, you know all about grey squirrels, a form of the eastern grey squirrel (*Sciurus carolinensis*), which patrol Stanley Park like hungry sentries.

In Toronto, large, black squirrels have moved in, bullied out the red squirrels and are now dominant. Black squirrels are actually a subgroup of the same species as the grey.

Grey and black aside, when we think of squirrels in Canada, it's usually the ubiquitous red squirrel (*Tamiasciurus hudsonicus*) that chirps angrily at us from every spruce tree. Smaller and faster than their larger grey and black cousins, they can be the both the most entertaining and the most destructive critters in the garden.

Grey form (above) and black form (below) of the eastern grey squirrel

LIFE CYCLE

While squirrels can technically live up to 10 years, they rarely get the chance. Normally, whether through disease, cold or predators, they're gone long before that.

Tree-dwelling species are usually solitary, so if you see them chasing each other around trees, it means they're either fighting or mating. Females have one to two litters of two to seven, naked, toothless, blind varmints annually.

DAMAGE

While cute to look at, squirrels can cause a lot of damage around the yard. They love to dig up and nibble on freshly planted bulbs and corms. Fruits, veggies, flower buds and even young bark can fall victim to their nattering little teeth.

Squirrels are infamous for menacing birds, and if your yard is invaded by squirrels, you will quickly find that the only birds left at your feeder are the magpies, crows and bluebirds that are strong enough to compete for the food you leave out.

If squirrels get into your attic, garage or anywhere else with live wiring, they have been known to nibble on wires enough to cause fires. If they move into these places, take immediate action to remove them, including calling a professional if necessary.

CONTROL

If you have large coniferous trees around your home, and especially if you live in an established neighbourhood where squirrels can leap from tree to tree, yard to yard, the pesky creatures are unavoidable. It's easy to buy a trap from the hardware store and trap them with peanut butter bait, but when you remove a squirrel, you're just creating a vacuum in the yard. A mature yard without squirrels is simply an open invitation to all the squirrels around it to move right on in.

"Squirrel-proof" bird feeders are very big business for people wanting to attract songbirds without having them scared away by bushy-tailed bullies. Many models simply don't work. Companies have tried everything from slippery tin sides to feeders that spin around when something heavy lands on them, but the success is mixed at best. Ask around for the best ones, read some reviews, and do your homework before buying.

Large evergreens and a garden full of treats will be paradise for a squirrel (above); they may be cute, but they are also bullies (below).

Voles and Mice

House mouse

I don't know about you, but for me, the spring thaw is one of the most anti-climactic events of the year. The first warm days of March, when the gutters fill with running water and winter's vice grip loosens a little, make me bristle with excitement. Once the snow is gone, however, and the yard looks like a flotsam-covered beach after a storm, it can feel like a long, brown road until summer.

A lot goes on under the winter snow. If you've noticed long, meandering lines that look like giant worms have crawled and chewed their way across your lawn, it's a good bet that mice or voles have been having a party down there.

IDENTIFICATION

While wild Canada has many species, there are really only three species of furry little varmints that we need to know about. The most common is the ubiquitous house mouse (*Mus musculus*). This industrious fellow is small, brown, agile and virtuosic at squeezing into our homes and lives.

The deer mouse (*Peromyscus maniculatus*) is larger than the house mouse, with large, round ears (think "Mickey") and a white belly. Deer mice are fairly rare in urban areas but are of special concern because they're a prime carrier of hantavirus.

Meadow voles (*Microtus pennsylvanicus*), which are often called by the

Deer mouse

misnomer field mice, are compact and stocky, with short legs and a short tail. They're the darkest-coloured varmint and are more closely related to muskrats than mice.

LIFE CYCLE

Under ideal conditions, house mice can have six to 10 litters throughout the year, each one consisting of four to eight young. The mice are sexually mature at six to eight weeks old. It doesn't take long for a mouse population to explode. Similarly, deer mice have several litters per year, though usually only during the warmer six months of the year.

Meadow voles are short-lived, notably monogamous and surprisingly prolific under their thick, snowy blankets. They have two main reproductive cycles per year, in spring and fall. The average litter size is four to eight young. Bad vole summers correlate to heavy snow-pack winters.

Meadow vole

DAMAGE

For such little creatures, mice are famous for giving us a big fright when they scurry across the floor. Mice in the home aren't only unsanitary and potentially destructive, but they're also a deeply unsettling reminder that our homes aren't as sealed off from the outside world as we like to think they are.

While mice are notorious for the damage they do to homes, voles don't climb and inflict their damage at ground level or, more insidiously, below. Mice look for convenient hiding spots underground, but voles construct elaborate tunnel systems, chewing their way through roots as they go.

Vole damage to lawns becomes glaringly evident as the snow melts. It will look like a 5 cm wide worm has squiggled and chewed its way across your lawn. You probably won't see much damage after this initial reveal; once spring sets in, voles move underground.

As summer advances, keep an eye on your trees. If some of them are struggling, with no evident disease, pest or other cause, you may have voles tunnelling around the roots (and thereby weakening them).

Mouse damage to lawns and gardens generally happens in winter. Mice burrow into the snow to look for old leaves, twigs, mulch or detritus to nest into, and they prefer deep, undisturbed snow for easy digging. Having established a nest, they next wreak havoc on your lawn and nearby plants by devouring any vegetation they can get their nasty little teeth around.

Watch for tracks scampering across the snow that sometimes vanish as they tunnel downward (look for a line the tail makes). If you had mice in the summer (and most yards do), then don't be fooled; they are down there.

VOLE CONTROL

Voles are so elusive that they can be difficult to control. If you want to protect your trees with a physical barrier, you'll need to wrap a metal mesh around them, just within the drip line to protect the roots. It will need to extend 30 cm above the surface and 15 cm beneath.

Like all rodents, voles hate capsaicin (the ingredient in hot peppers that gives them their spiciness). Blend some of the hottest peppers you can find with some onions, and drip the mixture into every vole hole you can find. Try to keep it off the surface in case dogs and cats get into it.

Deer mouse tracks in snow

Drip hot pepper juice into any vole holes you find in your yard.

As per mice, you can use snap traps to control vole numbers. Lay the traps at vole hole entrances and check them daily. As with capsaicin, place the traps just inside the holes to save the paws of local pets.

Badgers, coyotes, cats and especially owls adore fresh vole. Unless you live in a rural area, however, many of these critters can be hard to come by.

MOUSE CONTROL

Most information available about mouse control focuses on their summer behaviour. This is important, but because almost half of our year is winter and mice don't hibernate, it's worth asking what they are up to for all that time.

When it comes to surviving winter, think of mice as tender perennials. They have a better chance of overwintering if ample snow falls before the mercury plummets and if they have some mulch to snuggle into.

Mice will try to make themselves a houseguest throughout winter. Their mousy senses tell them that the house is full of warm carpets to curl up in and delicious couches to chew to pieces.

To protect your home, do a thorough clean-up around its perimeter in fall. Make sure that mulch, detritus and miscellaneous junk is scraped well away from the foundation. This will discourage varmints from making nests next to the home.

Don't abandon your backyard to be a wasteland of undisturbed snow.

Untouched, virgin snow fields may look serene, but they bustle with activity just below the surface. Calories are precious in winter, so mice will tunnel through the snow in the easiest direction. Get into your yard and stomp the snow down in a big circle around the house. Deny the mice easy tunnelling, and they will be less likely to scout for opportunities in your foundation.

When you're shovelling, try to heave extra snow onto your perennial and flower beds. On top of crushing mouse tunnels and nests, it will provide extra moisture for your plants in spring.

Don't put poison out in winter (I don't recommend putting it out at all). Unless they get into the house, mice don't breed in winter. All you need to do is discourage them from turning your winter lawn into a frozen buffet.

That being said, if they do get into your home: declare war. They can very hard to get rid of once they get established, especially if you don't have a cat.

Rake up your leaves in fall if you don't want to invite mice to make a nest in them.

Deer mice occur mostly in rural areas, but don't generally venture into houses.

Hantavirus

I want to take a moment to clear up some misconceptions about this dangerous disease. While it is potentially fatal and needs to be taken seriously, we tend to be more frightened than we need to be, as the vast majority of mice we encounter don't carry it.

If you see a house mouse, you don't have to worry; they aren't carriers. Coming into contact with their droppings, while admittedly disgusting, won't give you hantavirus.

The carriers are deer mice, which occur in rural areas and, thankfully, aren't known to venture into houses. If you have deer mice living near you, make sure to wear a respitory mask whenever you're near any evidence of them. Try to get rid of them as soon as you can, even if you have to hire a professional.

Heat Waves

A soaker hose keeps the water on the ground and means less evaporation.

When a heat wave hits, it's not just humans and pets who get uncomfortable. Prolonged high temperatures bring a lot of stress to gardens, but there are easy things that you can do (and avoid doing) to help your plants beat the heat.

TIPS TO MINIMIZE HEAT STRESS

Try to water first thing in the morning. It can seem like a chore, but a walk around the garden before heading to work can really clear the head. An early soaking will keep the soil cool as the mercury rises and give the roots a cushion of moisture into early afternoon.

If mornings aren't an option for you, the next best time to water is in the evening, after the scorch has soothed. However, if the nights are cool, mildew may set in.

Avoid overhead watering in the afternoon heat. It's a pet peeve of mine to see water being launched high into the air by sprinklers at 4:00 pm on a hot day. Not only is it highly wasteful—most of the water just ends up evaporating before it has a chance to soak into the ground—but it's expensive, as well. If you must water during the day's heat, invest in a soaker hose that keeps the water on the ground.

Plants have different types of root systems depending on where they come from. Native or other cool-climate plants tend to have deeper roots and are therefore better equipped to handle dry conditions. Tropical plants, such as tomatoes, peppers and many annuals, have roots close to the surface. Try to give them a little extra water; otherwise, when the soil dries and hardens, it could damage the roots.

Dark soil absorbs heat like a sponge, and hot soil can be dangerous to the roots just under the surface. Keep the soil cool by adding mulch around heat-sensitive plants such as salad greens, root crops and pansies. Cedar mulch or straw is best; avoid rock because it will heat up and actually make the problem worse.

Keep your lawn a little longer in the heat. Longer grass blades (around 7.5 cm) will help keep the roots cool by shading the soil, and more leafy tissue means more ability to transpire, which helps cool the plant.

Just like you don't want to go do push-ups all day in 30°C heat, your plants don't want to work hard in the heat, either. Avoid "surgical" tasks such as transplanting until the weather cools off a little. Root systems work overtime in heat and don't need any extra disturbances.

Don't do any transplanting during a heat wave; wait for a cooler, preferably cloudy day.

Organic mulch helps keep the soil cool.

Hail

I'm a prairie kid, and while I love the lushness of Vancouver and the grandeur of the St. Lawrence, I always come home to my big Alberta skies. Whether it's wide horizons of yellow canola fields or aurora borealis igniting long winter nights, our big sky country makes me feel lucky to live here.

There's a darker side to our big skies. When we see ominous-looking thunderheads rolling toward us like cavalry, we know it's time to batten down the hatches. When the thunderstorms move in, both in western Canada and across this vast country, one of the most destructive forces we have to deal with is hail.

DAMAGE

I've had my windows smashed to shards by hail, and my car covered in dings like someone beat it with a hammer. If that's what pelting hail does to glass and steel, imagine what it's doing to pumpkins and petunias.

There's rarely anything we can do to prevent hail damage. The storm usually comes out of nowhere and, when it does, it's not a good idea to run into the garden to cover plants.

Hail damages through blunt trauma, smashing and bruising until it beats plants down. As you survey the aftermath, you may feel like you've lost a lot of plants. The good news is that you probably haven't.

SURVIVAL ODDS

Whether a plant will survive a hail storm depends largely on its root system. With its above-ground leaves and branches being torn and broken, the plant will depend largely on its roots to supply it with the tools to grow back. The more established and hardier the plant, the better it will fare. If you planted it yesterday or it's a very tender plant with delicate roots to begin with, the survival odds aren't good if the storm was severe.

If it storms early in summer, when the plant is still growing at a feverish pitch, there's better chance of a full recovery. As soon as the tell-tale fall smell starts percolating into the air, the plant starts preparing itself for winter. Traumatized plants have a lower chance of surviving winter the later the storm hits.

If you're dealing with long-season fruit and vegetable crops, such as pumpkins or melons, I suggest tossing them if they have taken heavy damage. Depending on the time of year, they probably won't

Melons (below) probably won't survive a severe hailstorm, but your petunias (left) may surprise you with their hardiness.

be able to grow a fresh crop in time for harvest. Removing them early will leave nutrients in the soil and may improve next year's crop. Keep in mind that pockmarked fruits and vegetables are often still safe to consume, as long as bugs don't get into them.

RECOVERY

Whether or not you think they will survive, give them the chance to survive. Don't dig up any plants until they keel over because sometimes they will surprise us. Give them extra fertilizer to stimulate their roots and promote new growth to replace the battered leaves.

Hail damage is like shrapnel; it's messy. Tears and lacerations that don't heal are as dangerous to plants as they are to humans. If branches are torn, cut them off cleanly with sharp pruners.

Leaves can be tricky. While torn leaves are unsightly, especially on foliage plants such as hostas, they are also the plant's photosynthesis engine, which supplies nutrients that the roots will need to do the heavy lifting of creating new growth. If the plant is young and/or tender, keep as many leaves on as possible. The healthier and hardier the plant is, the more leaves it can afford to lose without losing its ability to recover.

The most important thing is to care for your plants. They are recovering trauma victims; treat them as such. Fertilize, water and otherwise baby them back to health. Expect them to be in shock for a few weeks and possibly to abort buds, flowers or developing fruit. In short, they need a break, but they may just bounce back better than ever.

If hail destroys your squash (above), pull it out; if your hostas (below) have been hit, keep as many leaves as possible.

Winter

Winters in Canada can, and often do, become downright epic. Between snow so deep you run out of places to shovel it and cold so intense the air freezes, it's a testament to human resilience how we're able to survive it.

But we must remember that while we have gortex jackets and central heating, our gardens remain pitifully exposed. Whether wind or snow or sleet or frigid temperatures, our intrepid plants must sustain it all.

A lot of people ask me which winter conditions are the most damaging to their perennials, shrubs and trees. My answers often surprise them.

SNOW

Winter's most infamous symptom can both protect and harm our plants. A considerable early snowpack, provided it's not oppressively wet and doesn't come by way of a violent blizzard, is the best thing that winter can offer. Snow is a splendid insulator, and if there's a 15 cm blanket of it in place before our first deep freeze, you'll find that many more tender plants will make it through than if the ground is bare.

The downside of snow is its often furious arrival. Early and late winter storms, with the wet snow they often generate, can doom otherwise healthy plants.

When the heavy snow arrives, slip on the boots and head outside to knock the accumulations off any laden, bending plants, especially evergreens, which catch snow in their needles. If you have well-pruned apple trees, which are trained to emphasize horizontal

If snow accumulates on your pine tree branches, gently knock it off (above); hilling up tender rose bushes in fall will protect their roots over winter (below).

branches, make sure to knock the snow off them (gently), as well. Use a long broom and stand well back, or you'll get a face full.

WIND AND SUN

Wind mercilessly desiccates everything all winter long, which makes it a real menace. Dry wind sucks the life out of every unprotected, tender branch it touches, turning them into husks incapable of leafing out.

If you have an older, established yard with big trees and ample vegetation on the perimeter, you shouldn't need additional wind screens. If your new yard is an open field, and especially if no fence has been built yet, I suggest protecting tender roses and cedars.

Mummy wrapping is a popular and very bad idea. Burlap pulled tight around vegetation prohibits air movement and can actually lead to rot during warm spells. The best idea is to hammer four stakes into the ground around the plant and create a simple burlap tent.

Sun plays much the same role as wind, in that it dries out exposed branches. The

burlap tent is the perfect protection. And as undisturbed snowy fields around your cedar act as a reflecting mirror when the sun is shining, stomp around in the snow to cut down on the damaging rays bouncing up from below.

UNEXPECTED THAW

Your plants will pass through winter best if they freeze and stay frozen. Freak February thaws can wreak havoc on tender root systems by causing the soil around them to frost heave, which disrupts soil just like it creates potholes in roads. There's little to do about this except to water your plants if it gets warm enough to thaw the soil.

WINTER PROTECTION

Before the ground freezes solid, water is the best protection we can offer our plants. Fall is often dry, particularly on the prairies, so break out the deep-water wand and the soaker hose, especially for tender perennials, roses, cedars, fruit trees and birch.

It seems counterintuitive, given how clammily cold we get while watering in October, but moisture hugs roots like a blanket and keeps them at a steady 0°C while dry soil dips much lower. Imagine your plants' roots as popsicle sticks, and it's your job to layer on the moisture so that, when the ground freezes, they remain happily insulated in ice. Doing so will be the best protection against anything winter throws at them, including all of the above perils.

Index

A

Agrilus planipennis, 26–28

Agropyron repens, 155–57

Ants, 76–79
  carpenter, 77
  field, 77
  mound. See field a.
  pharaoh, 77
  red harvester, 77
  thatch. See field a.
  wood. See field a.

Aphididae, 18–20

Aphids, 18–20

Apioplagiostoma populi, 121–23

Apiosporina morbosa, 115–17

Apis spp., 133

Apple maggot, 21–23

Araneae, 140–42

Araneus gemmoides, 141

Archips argyrospila, 39

B

Bees, 132–35
  bumble, 133
  honey, 133

Beetles
  See Colorado potato beetle
  See Crucifer flea beetle
  See Elm bark beetle
  See Emerald ash borer
  See Japanese beetle
  See Ladybugs
  See Scarlet lily beetle

Bellflower, creeping.
  See Creeping bellflower

Bindweed, field. See Field bindweed

Black knot, 115–17

Blight
  fire. See Fire blight
  late. See Late blight
  potato. See Late blight
  Typhula. See Snow mould

Bombus spp., 133

Borer, emerald ash. See Emerald ash borer

Bradysia spp., 84–86

Bronze leaf disease, 121–23

Butterfly, cabbage white. See Cabbage worm

C

Cabbage worm, 66–68

Caloptilia fraxinella, 37–38

Campanula rapunculoides, 150–52

Camponotus spp., 77

Canada thistle, 148–49

Cats, 166–67

Chickweed, 158–59

Choristoneura, 46–47
  fumiferana, 46
  occidentalis, 46

Cirsium arvense, 148–49

Coccidae, 90–92

Coccinellidae, 136–39

Colorado potato beetle, 61–63

Convolvulus arvensis, 153–54

Cottony psyllid, 24–25

Couch grass. See Quack grass

Crambidae, 40–42

Creeping bellflower, 150–52

Creeping thistle. See Canada thistle

Crucifer flea beetle, 64–65

Culicidae, 72–75

Cutworms, 69–71

About the Author

Rob Sproule knows that gardening can change the world.
As co-owner of Salisbury Greenhouse in Sherwood Park, Alberta,
he shares his passion for creative, healthy approaches to gardening
via newspaper columns, magazine articles, radio shows and books.
He loves talking to gardeners who are just starting out, and with
his unique blend of education and inspiration, he often gets them
hooked on growing.

Rob holds a Masters of Arts from the University of Alberta and
is an avid writer and reader. He lives in Sherwood Park with his
beloved wife, Meg, and their young son, Aidan.